International Perspectives in Sport Tourism Management

This timely book critically explores the past, present, and future trends and needs of sport tourism development from a global perspective.

Sports have become a major demand generator for many countries around the world, and consequently, the growth and status of this phenomenon is a major aspect of sport tourism. The volume includes case studies from all over the world, including industrializing countries and the Global South. Sporting events and sport tourism are two fast-growing sectors of tourism, and this book provides a current view of the interconnectedness of sport and tourism from behavioral, historical, economic, management, marketing, environmental, and policy perspectives. In doing so, issues, trends, and topics such as commercial, e-sport, emotional well-being, health, and cultural aspects of sports travelers are explored in detail.

This significant and insightful volume will be of pivotal interest to academics, students, and professionals in the field of sport tourism, as well as related areas, such as development studies, psychology, sociology, and geography.

Miklos Banhidi is a professor at the University of Physical Education and Sport Science in Budapest/ Hungary. His teaching areas are the Theory and Practice of Leisure and Recreation, Geography of Sport, Tourism and Recreation, and Project Development. His research interest is community development to improve citizens' quality of life, with a focus on how to design and manage healthy environments and organize powerful activities to achieve positive benefits. He has written numerous books and articles on recreation, geography of sport, health, and tourism. He has served as a visiting professor at the Karl Franzens University in Austria, at the University of Northern Iowa in the United States, and at Zhejiang University in China. He has given numerous lectures at several foreign partner universities. Along with his partners, he initiated and managed several student and youth exchange projects supported by the European Union. He served as the vice president of the United Games International, organizing festivals and workshops on peace and understanding worldwide. Currently, he is the president of the Sport Tourism Association within the Hungarian Sport Science Federation. Since 2009, he has been an elected member of the board of the World Leisure Organization and its Executive Committee.

Farhad Moghimehfar is a research chair in Sustainable Rural Tourism and professor of Recreation and Tourism Management at Vancouver Island University, British Columbia, Canada. His research interests are framed within the context of sustainable tourism, rural tourism, and nature-based recreation, focusing on community tourism and studies of human behavior and resilience of communities. Before joining Vancouver Island University, Farhad served as an instructor and postdoctoral researcher at the University of Alberta, Canada, and as an assistant professor at the University of Northern British Columbia, Canada. He has conducted research on community tourism and outdoor recreation and provided community and professional advisory services related to recreation, tourism, and park planning in Canada and internationally. He has served BC's communities by providing different services ranging from advanced data analysis to designing carrying capacity analysis systems. Farhad serves as an associate editor for the *Journal of Ecotourism* and has published in leading journals, such as *Tourism Management* and *Leisure Sciences*.

World Leisure Organization Book Series

Series Editors:

Dr. Marie Young
University of the Western Cape

Dr. Arianne Reis
Western Sydney University

This interdisciplinary Routledge book series aims to publish the most up-to-date and innovative critical perspectives on the leisure field.

The approach of the series corresponds to World Leisure Organisation's notion of leisure, viewing it as a multi-dimensional reality which includes tourism, sports, culture and recreation. The series also studies cross-discipline and cross-sectoral connections – within and beyond leisure studies and the leisure sector, as well as the development of useful educational tools for the leisure academic and professional communities.

In this series, leisure is acknowledged in its capability of generating alliances and a resilient ecosystem, which needs greater articulation and development of proactive actions, such as those discussed in this series. The series is thus deeply rooted in the theory-practice-policy connection and builds on previous works in the leisure field, as well as from the leisure sector, to further explore leisure's untapped potential for individual and societal well-being.

Book Series Publishing Partner
World Leisure Organization (WLO)
Arenal 1, 2ª planta
48005 Bilbao
Spain
Tel: 0034.946.056.151
secretariat@worldleisure.org; www.worldleisure.org

Titles in the series include:

International Perspectives in Sport Tourism Management
Edited by Miklos Banhidi and Farhad Moghimehfar

For more information about this series, please visit: https://www.routledge.com/World-Leisure-Organization-Series/book-series/Tourism.

International Perspectives in Sport Tourism Management

Edited by Miklos Banhidi
and Farhad Moghimehfar

Routledge
Taylor & Francis Group

LONDON AND NEW YORK

First published 2025
by Routledge
4 Park Square, Milton Park, Abingdon, Oxon OX14 4RN

and by Routledge
605 Third Avenue, New York, NY 10158

Routledge is an imprint of the Taylor & Francis Group, an informa business

© 2025 selection and editorial matter, Miklos Banhidi and Farhad Moghimehfar; individual chapters, the contributors

The right of Miklos Banhidi and Farhad Moghimehfar to be identified as the authors of the editorial material, and of the authors for their individual chapters, has been asserted in accordance with sections 77 and 78 of the Copyright, Designs and Patents Act 1988.

British Library Cataloguing-in-Publication Data
A catalogue record for this book is available from the British Library

ISBN: 978-1-032-76016-2 (hbk)
ISBN: 978-1-032-76017-9 (pbk)
ISBN: 978-1-003-47665-8 (ebk)

DOI: 10.4324/9781003476658

Typeset in Times New Roman
by codeMantra

This book is dedicated to academics, students, and professionals who are interested in getting involved in sport tourism development.

Contents

Figures

Tables

Contributors

Zoltán Ádám graduated as a molecular biologist from the Faculty of Sciences at the University of Pécs, Hungary, in 2018. Currently, he is doing his PhD program at the Institute of Pharmaceutical Biotechnology under the supervision of Dr. Krisztian Kvell and Professor Dr. Marta Wilhelm. His PhD research focuses on healthy aging of the human body and immune system. As a research fellow, he is part of the Wnt Signaling and the Food Biotechnology research group, where his job is to isolate and analyze miRNAs from different sources. In his free time, he does sports, reads books, or enjoys the tranquility of nature.

Barry Andrews is currently working as an associate professor, the Head of the Department, a project leader for High-Performance Sport, and a coordinator for the Sports Science program in the Department of Sport, Recreation and Exercise Science at the University of the Western Cape in Cape Town/South Africa. His teaching areas are adaptive physical activities, biomechanics, motor control, and high-performance sport. His research interest is in high-performance sport and maximizing the sporting potential of persons with disabilities, and he has written numerous articles in these fields. He has served as a visiting professor at Leuven University, Belgium, and has given lectures at several foreign partner universities.

Gıyasettin Demirhan is a faculty member at Hacettepe University, Faculty of Sport Sciences. He is the head of the Department of PETE, a board member of the Foundation for Global Community Health (GCH), a former dean of the Faculty of Sports Sciences, and a former president of the Turkish Sports Sciences Association. He is also a member of some international scientific associations such as AIESEP, FIEP, WLO, WPEA, and GCH. He is also a member of the Turkish National Olympic Committee. His research interests include critical thinking, instructional technology in physical education, the relationship between physical activity and academic performance and cognitive processes, and risk perception in outdoor adventure sports. He published 34 articles in international refereed journals and 50 articles in national refereed journals. In addition, he published two books and 15 book chapters in physical education and sports pedagogy. He presented more than 100 papers in international and national scientific congresses and has completed 14 international and national scientific projects in his field of research. He participated in many national and international scientific meetings as a keynote speaker and invited speaker. He also organized many international and national scientific meetings. He is a mountain guide and has long been interested in outdoor adventure tourism and sport, and risk and safety management. He has a book on Search and Rescue in Mountaineering, the only book on this subject in Turkish.

Mogamat Rida Desai has a BCom in Management and Industrial Psychology and a PG Dip in Disability Studies. He is a graduate of the Young African Leaders Initiative Graduate and is currently doing his master's degree in Sports for Development at the University of the Western Cape. He is a self-driven individual who needs to assist those in a less fortunate position. A horrific car accident in 2010 caused a spinal cord injury that has given him a different outlook on life. He is a disability activist with experience in many disability organizations and projects around inclusion and accessibility. His main aim is to make a positive change for future generations to live in better times.

Hannah R. Dudney is a first-year PhD student at the University of Toronto, Canada, in Rehabilitation Sciences with a focus on outdoor recreation for people with disabilities. After completing her undergraduate degree at Western University (London, Canada) in Sociology of Population, Health and Aging (2019), Hannah completed a diploma in Travel and Tourism (2021), with an interest in nature-based tourism. She continued to pursue inclusive outdoor leisure for all during her master's degree in Sustainable Leisure Management at Vancouver Island University (Nanaimo, Canada). Her thesis explored experiences of park crowding for individuals using mobility devices. She continues to develop her knowledge of park accessibility, accessible tourism, and employment for people with disabilities through various academic and governmental projects.

Bijen Filiz is an associate professor in the Department of Coaching Education at Kocatepe University, Turkey. Her areas of interest are instructional, curriculum, and hidden programs on physical education, teacher education, and sport education, sports tourism, and sports sociology. She published 27 articles in international refereed journals and 15 articles in national refereed journals. Also, she has eight book chapters in physical education and sport pedagogy. She presented 37 papers in international and national scientific congresses in her own research field. She is a member of some international scientific associations like GCHFL and TPSR Alliance. Also, she is a member of the Young People's Leisure Network and joins various camps in project scope. She joined the World Leisure Congress in Mobile, Alabama, in 2014 for the first book. At the end of this camp, they completed the book titled *Youth Leisure around the World* in 2015. For the second book, she joined the first camp and World Leisure Congress in Brazil in 2018 and the second camp in Hungary in 2019. She wrote a chapter titled *Safety and Leisure* into the second book. Through these programs, she knows new cultures and people, gains new experiences, and improves relationships.

Marcel Grooten is a lecturer at the Faculty of Leisure of Breda University of Applied Sciences (Netherlands). His teaching areas relate mainly to human capital, sports, and music events management. His research focus covers three different subjects: voluntourism in the marine conservation industry, authentic festivals, and events (e.g., Nadaam in Mongolia), and foremost the global development of professional cycling events. He has also worked in a diversity of leisure and tourism sectors (e.g., artists tour management, hospitality, tour guiding, and professional cycling). Marcel has also joined several grassroots volunteer projects in different continents. His interests are diving and cycling.

Brendon Knott is an associate professor in the Sport Management Department at the Cape Peninsula University of Technology in Cape Town, South Africa. His primary research interests are sports tourism, mega-events, legacy, and place branding. He gained his PhD (Bournemouth University, United Kingdom), focusing on nation branding as a legacy of the 2010 FIFA World Cup for South Africa. He continues to have influence and serve the sport tourism industry, most notably as a member of the Technical Committee of the 2010 FIFA World Cup in Cape Town, addressing the South African Parliament's Tourism Portfolio Committee, and chairing the 5th International Conference on Events (ICE 2021). He has conducted primary research at two FIFA World Cups and three Olympic Games. He serves as an associate editor for the *Journal of Leisure Research*, the Editorial Board of the *Journal of Destination Marketing and Management*, and the Advisory Board of *Event Management*, as well as conference co-chair of the Sports Leadership track at the International Leadership Association annual conference. He is passionate about the role of sport in emerging nations of the world, co-editing the 2021 book *Sport and Development in Emerging Nations*.

Jaroslav Kupr is currently working at Technical University of Liberec, Faculty of Sciences-Humanities and Pedagogy, Department of Physical Education, and at Jan Evangelista Purkyně University in Ústí nad Labem (also at the Department of Physical Education). His teaching areas are the theory and practice of leisure and recreation, non-traditional games, outdoor activities, and summer and winter courses. His research interest is physical activity and physical fitness, non-traditional games, and the development of aerobic fitness and outdoor activities. He is a member of Czech Mountain Leader Association (chairman of the Methodological and Security Commission 2014–2018), he has experience from the commercial sector (travel agency, skiing schools, centers, etc.), and he is the president of the Climbing Club (Lezci.com). His private focus is mountaineering (Mont Blanc, Mt. Elbrus, Pik Lenin, Mustag Ata, etc.) and cross-country skiing (Jiz50, Vassaloppet, etc.).

Klara Kuprova is currently working at Technical University of Liberec, Faculty of Sciences-Humanities and Pedagogy, Department of Physical Education. Her teaching areas are swimming, sport management, outdoor activities, downhill skiing, cross-country skiing, and mountaineering. Her research focus is on secular trends of school children, physical activity and physical fitness, and outdoor activities. She worked in the commercial sector (travel agency, skiing schools, etc.). She is a member of a climbing club (Lezci.com). Her interests are triathlon (World championship XTERRA – Hawaii) and mountaineering (Mont Blanc, etc.), trail running, and cross-country skiing.

Gyongyver Lacza is a recreation expert and a physical education and English teacher with more than 20 years of teaching and management experience from the active leisure field. She has an MSc degree from the World Leisure Organization's Center of Excellence and a PhD in Recreation. Since 2002, she has been a lecturer at the University of Physical Education Budapest, Hungary, and she has become the head of the Department of Recreation. Apart from her academic career, she has been involved in a number of leisure research and development projects on both national and international levels. Besides, she has been working at Develor, one of Hungary's leading training and consulting company, developing and leading soft-skill trainings for several years. She has a deep experience in organizing mass leisure sport events. In 2019, she was the leader from the university side in the Budapest European Capital of Sport project.

Tamas Laczko was born in Hungary in 1973, and he is a qualified economist and geographer. He obtained his PhD in 2009 in wellness tourism and his habilitation in 2021 in sports economics. Since 2002 he has been teaching at several faculties of the University of Pecs. He is currently an associate professor at the Faculty of Health Sciences of the University of Pécs. He is the author and co-author of textbooks on sport and health tourism and sport and society of four universities

His main research topics are the analysis of wellness tourism market trends in international and Hungarian context, and the development of the Hungarian wellness tourism market and the investigation of the demand conditions of passive sports tourism, with special emphasis on the consumption patterns of passive sports tourists.

In addition, he is a member of several research consortia that have investigated the returns, macroeconomic and social impacts of international sport events in Hungary. On these topics, he has published 82 scientific publications with 434 citations.

Rudolf Leber is an education and sports scientist, with a doctorate in philosophy, natural sciences, and sports science. He has studied Pedagogy, Special and Curative Education, Politics, Journalism and Communication Science, and Project Development. He graduated as a sports instructor, coach, and ski instructor. He teaches at universities in Central and Eastern Europe and Asia. The main focus of his scientific work is the development of the education and sports sector in urban areas.

Ping Ling is a full professor at the School of Physical Education and Health at Normal University in Hangzhou, China. His professional experience is member of the Board of Directors at the World Leisure Organization, vice president of Leisure Academic Association of Zhejiang, executive committee member of

China Sport Sociology Academic Association, professor and doctoral supervisor at the Zhejiang University, commissioner of National Educational Instructional Committee for Master of PE and Sport School of Health and Physical Education, Hangzhou Normal University, commissioner of National Sport Science Association Sport Sociology branch, and chairman of Table Tennis Association for Zhejiang University students. His research interests include leisure and recreation, sport industry, and sport marketing systems, as well as sports federation and sports club systems, sport management systems between countries and universities. He has had more than 142 research papers published in China and abroad.

Minkun Bill Liu is a professor and a PhD supervisor of Tourism Management at the Business School of Guangxi University, Nanning, China. His teaching and research cover tourism planning and development, the integration of culture and the tourism industry, and the economic benefits that tourism brings to destinations. His teaching and research interests include tourism industry convergence, wellness tourism, rural tourism, and all-for-one tourism. He is currently the assistant dean at the Business School and has undertaken several research projects at ministerial and provincial levels. Also, he has been funded by the National Natural Science Foundation of China twice. He has received numerous awards from industry and government for his work in the tourism discipline. He co-authored *All-for-One Tourism Era: The Study on the Upgrading and Development of Characteristic Tourism Counties in Guangxi* and won the third prize of Guangxi's 15th Social Science Outstanding Achievement Award. He was selected into the Excellence Scholars Program of Guangxi and the Training Program of Tourism Planning Talent in Ten Thousand Talents Plan launched by the National Tourism Administration of China.

Zinzan Magerman is currently an Honors student at the University of the Western Cape. He obtained his undergraduate Sport Science at the University of the Western Cape. His research topic was on the impact of the COVID-19 pandemic on university rugby players' participation in sport under the lockdown restrictions at a university in the Western Cape, South Africa. He presented a paper at the International Youth Scientific Conference on Leisure and Recreation with other authors. Further, he is working as an intern at Eye of the Tiger Rugby Academy, wearing several hats, namely, being a team manager, a rugby coach, and a strength and conditioning coach. He also assists with rugby coaching and strength and conditioning at his former high school, Excelsior High School. Furthermore, he is a youth leader at the Belhar Moravian Church and plays rugby for a local club, Tygerberg Rugby Club.

Angel Mahlalelal is currently a research intern funded by the Department of Science and Innovation-Human Research Council (DSI-HSRC), in the Department of Sport, Recreation and Exercise Science, Faculty of Community and Health Science, University of Western Cape. Her academic interests are in leisure education, sport management and administration, sport for development, and sports tourism. Under the mentorship of Makhaya Malema, a member of the World Leisure Organization Young Ambassadors Network, and Professor Marie Young, a member of the World Leisure Board of Directors, it is her middle-term career goal to obtain a master's degree in the field of sport and recreation upon completion of her internship. She is passionate about health and helping people increase their quality of life by promoting an active lifestyle. During strict lockdown alert levels in our country, she engaged with the Department of Health in Mpumalanga Province, educating and raising awareness about the importance and benefits of recreation and exercise. She is currently engaged with a non-profit organization in the Western Cape Province, assisting people with disabilities to improve their quality of life through sport and physical activities.

Lawal M. Marafa is a professor at the Department of Geography and Resource Management, The Chinese University of Hong Kong. He is also the director of the Postgraduate Program in Sustainable Tourism. He was awarded the "Exemplary Teacher Award" in 2005, and he received the Royal Belum Inaugural Ecotourism Award, Perak, Malaysia, in 2007. He served as a facilitator at the Clinton Global Initiative meeting in Hong Kong in 2008. He serves as the vice chair of the Board of Directors of the World Leisure Organization. In the 1980s and early 1990s, he worked with various organizations in Nigeria, including as a principal land use planning officer at the World Bank, assisted Sokoto Agriculture and Rural Development Authority, and worked as a principal forest planning officer, Federal Ministry of Agriculture and Natural Resources, Federal Department of Forestry, Abuja.

Although residing in Hong Kong, Dr. Marafa acts as a bridge between Africa and Asia. He serves as the regional coordinator for the New Partnership for Africa's Development, Business Group (NEPAD-BG), and also coordinates the African Business Roundtable in Asia. He attended the African Development Bank Board of Governors meeting in Shanghai in 2007, and he also attended the African World Business Congress, Arusha – Tanzania, in 2007. He has served as a resource person and keynote/invited speaker at the TICAD-UNDP-World Bank Fifth Africa-Asia Business Forum for Sustainable Tourism (AABF V) hosted in Kampala, Uganda, in 2009. His teaching and research interests cover tourism, environment and climate change, ecotourism, leisure, recreation planning and management, agriculture and natural resource management, and sustainable development. Over the years, Dr. Marafa has published over 70 articles in various outlets and presented at about 30 international conferences.

Lenia Marques is an assistant professor of Cultural Organizations and Management at the Erasmus University Rotterdam (the Netherlands). Previously, she worked as a researcher and lecturer in different countries (Portugal, United Kingdom, and the Netherlands). She served as a member of the Board of Directors of the World Leisure Organization between 2016 and 2021. She has published extensively in cultural tourism, creative industries, and cultural management. She is the co-editor of *Leisure and Innovation*, a special issue of the

World Leisure Journal (2019); *Event Design: Social Perspectives and Practices* (2015); *Intercultural Crossings: Conflict, Memory and Identity* (2012); *Exploring Creative Tourism* (2012), a special issue of the *Journal of Tourism Consumption and Practice*. She has also been advising different countries, regions, and cities on culture and tourism policies.

Petra Mayer is a physiotherapist (BSc) and a recreation expert (MSc). As a physiotherapist, she worked in the home care field for years, having previous experience in supporting a variety of patients, ranging from children suffering from developmental problems to adults and the elderly affected by and recovering from injuries and movement disorders. Now she is working as an assistant lecturer at the University of Pecs in the Institute of Sport Sciences and Physical Education, teaching anatomy, physiotherapy, and recreation courses to students of sport sciences. She is currently doing her PhD studies in biomechanics.

Kinga Nagy is currently working at the Department of Recreation at Hungarian University of Physical Education and Sports Science in Budapest, Hungary. As an assistant lecturer, she is teaching sport tourism, health tourism, recreational watersports, winter sports, experiential learning, etc. to students of sport sciences. Her research interest includes adventure tourism participation, flow as a motivation for continued participation in leisure sports, and segmentation of active sport tourists. She has practical experience in the commercial sector (front office manager in a Destination Management Office, Watersports Center manager, and instructor). She is currently doing her PhD studies in sport tourism.

Tatum-Leigh Petersen graduated with an Honors degree from the University of the Western Cape in 2021. She is passionate about recreation and sports development in Third World countries. Her research topic was on the impact of the COVID-19 pandemic on university rugby players' participation in sport under the lockdown restrictions at a university in the Western Cape, South Africa. She presented a paper at the International Youth Scientific Conference on Leisure and Recreation with other authors. Tatum contributed as a co-author to book chapters related to sports coaching and tourism for persons with disabilities. Additionally, she participated in an institutional project for children aged ten and younger, assisting them in developing gross motor skills through participation in various activities with their peers.

Donna-Leigh Reid graduated with an Honors degree from the University of the Western Cape in 2021. She is passionate about Recreation and Sports Development in Third World countries. Her research topic was on the impact of the COVID-19 pandemic on university rugby players' participation in sport under the lockdown restrictions at a university in the Western Cape, South Africa. She presented a paper at the International Youth Scientific Conference on Leisure and Recreation with other authors. She contributed as a co-author to book chapters related to sports coaching and tourism for persons with disabilities.

Monir Shahzeidi is a PhD candidate at the University of British Columbia (UBC), Canada. Her doctoral research focuses on the Influence of Leisure on the quality of life of women refugees. She also investigates people's behavior during outdoor recreation activities and parks and protected areas carrying capacity. Prior to this, she completed her MA in Sustainable Leisure Management at Vancouver Island University and worked as a researcher at the University of Alberta where she studied campers' pro-environmental behavior and UNESCO World Heritage Site visitation behaviors. She has several years of experience as a senior accountant and pursued Chartered Professional Accounting before she joined VIU. Her passion is to understand refugees' challenges in order to improve their quality of life. She is also interested in the role of nature-based recreation in the well-being of people's life.

Chiung-Tzu Lucetta Tsai is a professor of Leisure and Sport Management in the Business School Department of National Taipei University in Taiwan. She is also the president of the Chinese Taipei Waterski and Wakeboard Federation and the Taiwan Leisure Association. Moreover, she has been serving as a member of the Board of Directors for the World Leisure Organization from 2012 to 2017 and a commissioner in the Gender Equality Committee and the Ocean Affairs of the Executive Yuan from 2014 to 2016 and 2018 to present. She has granted the "Academic Research Award" in 2009, 2010, 2012, 2013, 2019, and 2020 in Taiwan. To date, she serves on a number of journal reviewers, including the *World Leisure Journal* (SCOPUS), *Leisure Studies* (SSCI), *Sociology of Sport Journal* (SSCI), *International Review for the Sociology of Sport* (SSCI), *Journal of Research in Education Sciences* (TSSCI), *International Journal of the History of Sport* (SSCI), *Social Indicators Research* (SSCI), *Journal of Leisure Research* (SSCI), and *Tourism Management* (SSCI). She is the editor-in-chief of *International Leisure Review* and *Associate Editor of Leisure Sciences* (SSCI). Dr. She has written on the topic of women in leisure and sport. Her research has appeared in the *International Journal of the History of Sport* (SSCI), *Social Indicators Research* (SSCI), *Leisure Studies* (SSCI), *World Leisure Journal* (SCOPUS), *Sport, Education and Society* (SSCI), *Leisure Sciences* (SSCI), *Tourism Analysis* (SCOPUS), *International Journal of Business and Information* (ABI), *Contemporary Management Research* (ABI), *International Journal of Business and Administration* (ABI), and *Ecoforum* (EconLit). In the past 15 years, she has received 60 national grants, up to USD 600,000, to examine leisure and health issues among women.

Yolanda van der Westhuizen qualified in 2009 as an occupational therapist and is currently working in the Department of Health with a specific focus to establish identity among patients living with TB and HIV. With numerous publications, her latest is set to be published on her master's degree, elaborating on the phenomenology of traveling for people with disabilities. In 2019, she established an online coaching program, Xplorable, aimed at people with disabilities. Here she facilitates clients to discover their new self, based on experiential learning and their sensory profile.

Marta Wilhelm was born in Hungary in a happy family as the daughter of a kindergarten teacher and an economist and grew up in a big village close to the Danube. This beautiful natural environment and her grandfather made it possible and basically compulsory to grow up as a person interested in nature. She became a biologist, but as a kid she was already reading hundreds of books about nature and species on Earth, chose an idol, Jane Goodall, and that did not change during her life. Entering the University of Szeged, she had to change her future plans (earlier she wanted to become an ethologist. For her master's thesis, she worked with alcohol-preferring rats, and after the master's degree, she worked for the Medical University in pharmacological and behavioral sciences. Her PhD thesis was based on retinal structures of lizards and frogs, working also for Flinders University, Adelaide, Australia. After getting a postdoc position in the United States (from 1997 to 1999, Columbia University, New York), she was changing her scientific interest, working in the field of neuro-immunology. Since returning back to Hungary (1999–), she works for the University of Pécs, Faculty of Sciences, Institute of Sport Sciences and Physical Education, teaching courses related to Exercise Physiology, fitness testing, healthy lifestyle, conducting many scientific projects with her students related to healthy lifestyle, healthy aging, recreation, and health.

Marie Young is an associate professor in the Department of Sport Recreation and Exercise Science, Faculty of Community and Health Sciences, University of the Western Cape, South Africa. She lectures in sport and recreation management at the undergraduate and postgraduate levels. Furthermore, she lectures on short courses as part of community engagement projects. Her research interest lies within leisure and recreation, more specifically programming, and therapeutic recreation to promote quality of life, health and wellness, youth development, disability, and changing environments making communities livable. However, conducting research and supervision in sport and recreation requires an individual to be diverse and supervise postgraduate students across disciplines in the faculty. The field's diversity leads to her involvement in various interdepartmental and inter-institutional projects nationally and internationally, serving as leader of these projects. She was also a visiting professor at the University of Northern Iowa, Winston Salem State University, and East Carolina University in the United States as well as Karel de Grote University of Applied Sciences, Belgium. Marie presented papers at conferences and submitted collaborative grant applications that have been successful. Internationally, she serves on the Board of Directors and Executive Board for the World Leisure Organization and the Editorial Board for the *World Leisure Journal*. Her membership on these boards promotes her academic and professional profile. Over time, she developed professionally in project management and internationalization. It includes taking leadership in projects, especially international projects, leaving a footprint beyond South Africa through her international engagements, serving on committees, and making links with international institutions for collaboration in learning and teaching and research activities.

Yingqi Zhang acquired her BA in English from Capital Normal University, then she went to the Hangzhou Normal University to pursue her master's degree, where she spent four years studying the management of the leisure industry. During the postgraduate period, she participated in the "Future Leader Program" of the World Leisure Conference in South Korea. She now works as Interest Rate Derivatives senior manager at Tullett Prebon SITICO (China) Limited, with over ten years of experience as a money broker, focusing on providing market quotes for interest rate derivatives trading.

Lijun Jane Zhou is a full professor and deputy dean of the College of Education, Zhejiang University, China. She teaches the courses of Sports Management and Sport Sociology for undergraduates, Theory and Practice of Leisure for master and doctoral students. Currently, her lines of research are focused on cross-cultural analysis on leisure behavior and leisure project planning. She has published more than 50 academic articles on peer-reviewed journals, two monographs, and several chapters as well. In the recent five years, she has been the chief leader to finish the national, provincial, and local governmental research projects on sports leisure programming. The total funding of the program is over 3 million RMB. Some of her research achievements were adopted by the provincial and municipal governments in China.

Foreword

Sport tourism is framed within the broader sector of leisure, in which it competes with other subsectors and products intended to occupy people's free time. Because of this, it is important to understand leisure as a whole since in recent years, we have seen that the ways in which people think about leisure have changed and evolved. We can say that leisure is everything that is freely chosen, without a utilitarian purpose but, fundamentally, because it has intrinsic satisfaction and enjoyment. However, we know it can be much more.

We are at a point that asks us to consider the importance of leisure, as a field that touches many sectors and ecosystems: cultural, sports, tourism, or recreation. At this point, it is important to understand that there is a distinct role played by the consumer of leisure compared with the role played by leisure operators, academics, or policymakers.

The first key of this point is in the context and in what the crisis means: it has been verified, in a reliable way, that leisure is fundamental for society. However, this need is rarely translated. In relation to this, one of the new developments of leisure is that societies, in terms of leisure, are entering in the co-creation of value. Within this reflection, there is a part that has to do with the fundamental right to leisure, guaranteed in articles 24 and 27 of the Universal Declaration of Human Rights and in other political documents of global scope, such as the 2030 Agenda for Sustainable Development. As it is mentioned in the WL Charter, provisions for leisure for the quality of life are as important as those for health and education.

This means that leisure, as a democratic practice, is a crucial and powerful instrument for promoting social development, and at the same time, it plays an important role in the competitiveness of the economy, especially in the third sector, being a source of innovation and creation and consolidating itself as a tool to improve the lives of people and economic and social transformation. Some developments are related to the establishment of appropriate mechanisms to guarantee it and recognize the intrinsic and, at the same time, diffuse value that leisure implies, because, sometimes, the lack of consensus on the valuation of the elements of leisure can cause that intrinsic value to be lost. In this sense, for example, the World Leisure Organization (WLO) has proceeded to review one of its great contributions: the WL Charter for Leisure 2020, in order to cover important updates and changes in line with emerging social and global problems.

Another part of the reflection, derived from that fundamental right to leisure, is creation and enjoyment. This has to do with the sector itself and with its professionals, as we must ensure that we really identify with who we are. It is necessary to invest in research and development (R&D) with a social base, without losing sight of its relationship with science and innovation.

Leisure has become in recent years a strategic sector due to the important role it plays as a catalyst for economic activity and employment, thanks to its creative and innovative component. The activity stop has been used to create new ideas and products. Here creativity plays a fundamental role and, of course, digitization and innovation.

There is no doubt that, for some time now, the support par excellence of knowledge is digital support. Among the advantages of digitization, we find that it allows us to do networking in a talented environment. It also has an impact on the labor market as it facilitates, for instance, access to culture, leading to greater social and professional inclusion. This digitization has also changed the creation of leisure products and services, events, audience engagement, marketing, and in the exhibition and distribution of leisure products. This quantitative progression pushes us to a profound qualitative change. In a society in which knowledge and skills become obsolete before even becoming routine, practically all members of society become part of the creative process of the leisure field, perhaps not generating new ideas, but learning and incorporating them into their lives.

The objective of the new leisure must be to provide people with high-quality, rewarding, and inclusive leisure activities that promote self-management of time and try to avoid isolation, since the value of it is about making leisure more equal and accessible. As an international human right, it has been confirmed that there is a relationship between leisure and learning throughout life, also called Life Long Learning (LLL). For all this, the leisure field, understood as a network that is capable of generating alliances and a resilient ecosystem, will need greater articulation and development of proactive actions, such as the ones discussed in this book.

In this framework, we have to bear in mind the importance of cities, which are configured as the first level where human interaction occurs, subsisting as places and spaces of influence and innovation in an increasingly globalized world. Additionally, it is a fact that the world is facing a process of demographic change, and its population continues to increase. Indeed, according to the latest data provided by the United Nations, it is estimated that the world population could grow to around 8,500 million in 2030, 9,700 million in 2050, and 10,900 million in 2100, concentrating, for the most part, in cities. Thus, the demographic and industrial transformation, and the consequent acceleration of investments lead cities to be so significant today. Here, the sport tourism management, understood as a technique to respond to the challenges posed by a dynamic and changing environment, will be essential, as readers can interpret by reading both chapters regarding sport tourism and the role of open spaces in cities, and managing sport tourism in communities, which demonstrates how Chinese industry has become one of the most favorite tourism destinations in the world.

Focusing now on sport tourism itself, it should be pointed out that its specificities of consumption are characterized, in general terms, by being more than a commercial transaction, since human values are put into play. Also, they transcend personal interests, since aspects such as learning, enjoyment, and self-realization move us.

Over and above, it becomes crucial to understand that one of its characteristics is the different range of elements that compose it, which, as a result, allow the development of individuals and society or territory. In general terms, these elements could be grouped into:

Personal: Access and enjoyment in all its versions have an effect on people's development at the same time they improve life quality, related to human benefits, specified in one chapter of the book, such as physical, mental, social, or learning outcomes.

Social: Sport tourism can work on issues, such as respect and tolerance, social cohesion, inclusion, the promotion of citizen participation, the protection of human rights, the gender approach, or the transmission of values.

Education: In the sport tourism field, the human factor that works in the sector is often the key factor of success for companies and production centers. Sustainable growth will depend largely on the sector workforce and on its rapid adaptation to the current challenge. The correct approach to sport tourism education will allow it to respond to the needs of both the private and public sectors, since as mentioned in the chapter on winter sport tourism, experienced educators are needed.

Territorial: The growth of sport tourism positively affects the creation of resources, the generation of employment, and the projection of the destination image, which directly and indirectly affects the social improvement of the environment. This element allows the creation of parallel projects in different fields or sectors, such as cultural, educational, or, even, gastronomic.

Socioeconomic: Sport tourism is a strategic asset for socioeconomic development. As explained in the book, both of them are big industries, which made the perfect marriage.

Cultural: This refers to the expression of communities and cultural diversity, as we can see in Part II (Sport Tourism Destination Management) in the chapter dedicated to sport traditions, habits, rules, and policies – A case study from Turkey.

Political-Institutional: Sport tourism impacts on the strengthening of the institutions, the cooperation between them, and the coordination in different projects.

This new period has shown us that despite all the scientific knowledge, we have not been able to understand the human experience in its full potential and capacity for successful or unsuccessful adaptations, mainly because adverse events have been developed in an accelerating way, which means that the rapid responses taking into account have not been planned and examined in a quiet manner. Additionally, sport tourism and leisure as a whole are quickly evolving, and their contents, to some extent, are prevailing upon their material substratum. It is an expansive process, requiring a balanced combination of general analysis and specific focuses. That is why research, such as the one mentioned in the book of the case study for Hungary, policy guidance, and cooperation are more needed than never. It will be necessary to rethink some aspects, especially in what concerns planification and access. We will need innovative actions. More data should be used, so as to make sense of the complex environments we live in, developing solution-based approaches and implementing policies to influence and drive better outcomes.

Of course, all these mentioned actions should be delivered in a more sustainable, environmentally friendly, and accessible to all way. Over the past half-century, there has been a growing call to action to rethink how societies persist. One of the earliest and most cited global documents addressing this issue is the Brundtland Commission Report, which proposed an imminent need for sustainable development. Research and development initiatives are rarely funded these days without direct mention and measurement of sustainability impacts, even though "sustainability" as a term is ambiguous and its operability is continually discussed. Sustainability principles refer to the environmental, economic, and socio-cultural aspects of tourism development, and a suitable balance must be established between these three dimensions to guarantee its long-term sustainability.

Considering all the aforementioned, we should consider that one of the keys to sustainability in leisure, and in consequence in sport tourism, is the need to identify and question how leisure or sport tourism is part of the problem and part of the solution, proposing and monitoring new approaches that improve sustainability through leisure or, in this case, to sport tourism. In consequence, these fields need greater articulation and that we be proactive, understanding the field as a network that is capable of generating alliances and a resilient ecosystem. Global engagement plays a crucial role here. For example, WLO is working on a new WL Cities Initiative, which aims to support cities in the development of policies, actions, and programs in the field of leisure, particularly in the face of ongoing changes and mutations.

Another example is the WL Games event. Being aware that sport and physical activity have the power to change our lives, as it is enshrined in UNESCO's International Charter of Physical Education, Physical Activity and Sport, WLO decided to create WL Games event, with the aim to use sport as a vital, powerful, and cost-effective tool to inspire and to unite people. All sports are welcome in the WL Games, since every sport can be used to deliver a positive social impact. The WL Games are hosted in communities that reflect and/or have the desire to develop sustainable leisure-oriented environments and are thought of as a sport for all competition, a community festival, a way of promoting the cultural heritage of a community or country, or as an end-destination tourist attraction that focuses attention on the hosting community.

Because all of these mentioned before, I strongly recommend to read this book, since it gives a worldwide approach of sport tourism management, which will help the reader to understand the importance of both tourism and sport, their relationship, and the need to act in a proper way at the time of operating in this subsector, thanks to the management activities and study cases proposed, since Aristotle quoted: Men acquire a particular quality by constantly acting in a particular way.

Cristina Ortega, COO, World Leisure Organization
Joanne Schroeder, Chair, World Leisure Organization

Part I

Trends in Sport Tourism Management

Photo: Jaroslav Kupr

1 Sport Tourism from Researchers' Perspectives

Miklos Banhidi

Introduction

Sport and tourism are two "mega-industries" of our society, mobilizing numerous industrial subsectors. The intersection of these two areas resulted in the emergence and development of sport tourism (Dreyer, 2002; Gibson, 1998; Röthig, 1992).

According to World Tourism Organization data (2008), the number of tourists has reached 1.6 billion by 2020, which has doubled in 20 years. The leading continent is Europe which hosts almost 50% of all international travellers. However, in the future, the role of Africa and Asia may also increase significantly behind North America.

Tourism affects 12% of the world economy directly or indirectly. Similarly, dynamic development in the field of sport tourism has been registered by tourism researchers, whose increasing supply is among the leading products of tourism (Pigeassou, 2004; Turco et al., 2002). In South Africa, 30% of the tourism sectors have been registered in favour of sports tourism (South African Government Information, 2007). The demand for sports-related travel is very well illustrated by some surveys. According to a survey commissioned by the Canadian Department of Tourism, 49.1% of American and 37.2% of Canadian travellers are interested in sporting events (Lang Research Group, 2004). According to statistics, the world's most visited sporting event, the Tour de France, is visited by 15 million people every year, 40% of whom come from abroad (Lamont & McKay, 2012). The Olympic Games have attracted 5 million visitors in Sydney, 3.5 million to Athens, 7 million in Beijing and around 8 million in London and will continue to be considered as a major sport tourism attraction.

According to a survey of the 1996 Olympic Games, 92.5% of the participants had never been to the Olympics before, but the visitors did not want to miss this opportunity (Neirotti et al., 2001). When Beijing was nominated to host the Olympic Games, the hotel capacity was increased by an average of 7% per year, reaching 80,000 rooms, three times more than four years before in Athens. For the 2010 Vancouver Winter Olympics, a major tourism campaign was launched in 2006 (Simpson, 2005). After hosting the 2010 tournament, South Africa's international tourist arrivals grew at an annual average rate of 7.4% in the three years to 2013, when it received 9.6 million foreign visitors (Fletcher, 2016). The London 2012

DOI: 10.4324/9781003476658-2

Table 1.1 Sporting Habits of Austrian and German Citizens

	Free time (%)	During travel (%)
Hiking	45	70
Bike touring	60	20
Mountain biking	15	5
Inline skating	10	10
Mountain climbing	10	10

Source: Brämer (2002).

Organising Committee has also declared the Olympics as the biggest tourist attraction of the country's history. Brazil spent $11 billion on hosting the World Cup in 2014 (Panja, 2014).

In order to encourage active tourism, New Zealand has expanded its range of active sporting activities in the natural environment, which has doubled the number of guests (Higham et al., 2006). In Germany, there are 10 million regular hikers, who average 360 km per year and 20 million who hike occasionally (Brämer, 2002). In the United States in 2003, 204.8 million people went skiing and 177 million went hiking in the mountains (USDA, 2003). Tourists were predominately middle-aged, affluent males who take an average of five short-break trips annually of about 400 miles per trip during the spring and summer months, and in the process spend approximately US$400 per trip (Buning et al., 2019) (Table 1.1).

Types and Characteristics of Sport Tourism

Both sectors, tourism and sport, are characterized by factors that are clearly linked to:

* Quality of life – In both areas, the quality of lifestyle is an important factor, as tourists want to live their life even on a higher level than at home.
* Relaxation – A significant percentage of travellers leave their homes and look for resorts to relax. The situation is similar for those seeking recreational sports who are looking for places and ways of spending their leisure time in order to balance their daily activities.
* Travel – Tourism is based on travel, reaching destinations in different ways.
* Sport activity – A physical activity that requires psychomotor skills, many of which have evolved in the past based on different locations, devices and tests.

Based on the above, sports tourism can be defined as sport travel, during which a tourist uses tourism services (travel, accommodation, meals…).

There are several forms of sport tourism (Figure 1.1):

Event Sport Tourism

Many sports fans travel to attend sporting events as spectators. It is mostly connected to tourist services, such as visiting the host town and staying there overnight.

Figure 1.1 Sport Tourism Categories.

Tournaments designed to attract sport participants to a destination have become significant elements of tourism marketing (Green & Chalip, 1998).

Participative (Active) Sport Tourism

Most sport lovers travel from their place of residence to another to experience sporting activities (Getz, 1998). The main goal is for them to practice their favourite activities in a different location, like *playing with space* (Geffroy, 2017) and enjoying the environmental benefits.

Educational Sport Tourism

This subsection includes school-organized excursions, summer camps and training camps. According to Getz and McConnell (2011), many sport tourists are highly involved in competitive mountain biking and are primarily motivated by self-development through meeting a challenge. The tourists participating in educational sport tourism can be viewed as serious sport tourists by applying serious leisure and ego involvement theory.

Cultural Sport Tourism

The cultural experience in sport can strongly motivate travel (Funk & Bruun, 2007). Some sporting events can serve as cultural tourist attractions that can facilitate authentic experiences (Higham, 2018).

Commercial Sport Tourism

In this sport sector, sports industry areas will be used to attract tourists and earn profits. It has been suggested that if sport shopping can attract travellers, they can be called sport tourists. We believe this is accurate. Even the Tripadvisor website (www.tripadvisor.com) recommends trips to sport outlets in New York (Jersey Gardens), London (London Designer Outlet) or Hong Kong (Citygate). Also, Hungary's largest sport outlet centre near Budapest (Premier Outlet) attracts more than 2 million Hungarian and foreign visitors every year on an area of 22,900 m². This outlet centre accommodates hundreds of premium fashion, lifestyle and sport brands, offering standard 30–70% discounts every day of the year.

Sport Expo Tourism

In the last years many sport exhibitions were organized, which attracted many tourists. Most of the sport equipment producers or distribution companies have interest to market their products in national or international expositions (Jisha et al., 2004). Destinations such as Cologne in Germany have become internationally known for sport expos combined with conferences. Many other cities have followed the example, such as South Africa and Australia (Saayman, 2012). The main idea of hosting an expo is that it is a marketing campaign not only for the products but also for the host destinations. In the US, Colorado, Sacramento, Salt Lake City, Denver and Scottsdale host large events called Sportmen's Expos once a year, which brings together enthusiasts of outdoor activities – camping, fishing, off-roading, hunting, etc. There are more than 1,600 companies that introduce their products at free theatre gear testing areas, and contest venues. One of the youngest but fast-growing events in this sector is the China Sports Culture Expo (CSCTE), which takes place in Guangzhou, China, supported by the Chinese Olympic Committee and Administration of Sport of China. The Sport Tourism Expo on a 40,000 m² exhibition area attracts more than 450 producers and 2,000 media companies. According to Eventseye.com (2022), across the world 446 trade shows are planned related to sports and tourism.

Therapeutic Sport Tourism

Therapeutic sport tourism is a new form in the sport tourism sector, where sport activities can be used for therapeutic purposes. In Canada, for example, dragon boat activities are organized as an additional treatment for cancer patients (Harris, 2012). For tourists who are travelling for health purposes, sports therapy is one of the aspects of healthcare that is specifically concerned with the prevention of injury and the rehabilitation of the patient back to optimum levels of functional, occupational and sports-specific fitness, regardless of age and ability (Ardern et al., 2016; Banhidi, 2004a). It utilizes the principles of sport and exercise science, incorporating physiological and pathological processes to prepare the participant for training, competition and where applicable, work (Zachazewski and Magee, 2012). All over the world, spa communities and health resorts were built (in Europe around 1,400), which offer several sport activities, many of them for therapeutic purposes. The importance of this subsection is growing fluently, with an annual turnover of 45 billion euro and 180 million overnight stays.

E-Sport Tourism

E-sport is a quick growing business, which attracts many travellers (Dilek, 2019). "E-sport" is a term that signifies the seamless interpenetration of media content, sport and networked information and communications technologies (Hutchins, 2008). The e-sports market is experiencing rapid growth. By 2025, the market is expected to generate over $1.87 billion USD in revenue (Statista, 2025). According to DiFrancisco-Donoghue and Balentine (2018), competitive video gaming has a global audience of more than 320 million, and the number of viewers who regularly follow the industry and tune in to watch international competitions is projected to reach 589 million by 2020. E-sport events, such as the World Cyber Games (WCG), constitute an important attraction for the tourism sector (Hamari & Sjöblom, 2017). The Asian Games in 2018 officially included six e-sports: Arena of Valor, Pro Evolution Soccer, League of Legends, Hearthstone, Starcraft II and Clash Royale, which are included in the overall medals. Those are being included in the 2022 Asian Games too.

In practice, however, these forms are often mixed. For example, tennis players coming to Wimbledon are welcome to watch the current matches, to try the local tennis courts and to visit the local museum. Tour de France organizers also offer cycling facilities after the race.

Both sports and tourism science research have been booming for decades, but more research projects should investigate relationships between sport tourism theory and different scientific disciplines.

The relationship between geography and tourism has been regularly studied by geographers since the 1930s (Carlson, 1938; Jones, 1933; Selke, 1936). Already, in 1935, Brown made a captivating statement that tourism research is a promising unexplored area, providing an opportunity for a more detailed description of new geographical areas. Since the 1960s, regional tourism studies have come to the forefront (Wolfe, 1967), as a result of which the economic rise of some regions and the development of leisure culture have been paralleled with the development of tourism.

Economists have sought to understand the economic effects of tourism as a function of supply and demand (Crisler & Hunt, 1952; Deasy & Griess, 1966). Their research results have shown that the choice of tourism products is significantly influenced by price factors, among which sports tourism is one of the cheapest. This is true despite the fact that building a sports infrastructure or organizing sporting events involves a significant budget. Due to the large number of participants in sports, they have become a major tourist attraction, and as a result, the return on investment is generally high. According to a case study, in 1994 the economic impact of the World Cup in Dallas, Texas, was estimated at $301 million, double the Houston Livestock Rodeo at $150 million and six times the most popular Mobile Cotton Bow at $56.5 million (Kay & Jackson, 1991).

In their interpretation of tourism from a psychological point of view, researchers sought to discover why people participate in sports. According to sports psychologists, this can be traced back to the two main components of sport motivation, called the need motivation and the goal motivation (Woods, 2004), which drives people to practice and to develop their motor skills. According to these, sports tourists, whose lifestyle includes daily exercise or sports activities, are looking for sports tourism offerings based on general motivations. At the same time, besides the need for development and interpersonal relationships, adventure seeking and competition play a significant role (Bang & Eaton, 2006).

Since the 1980s, several models have been developed to describe sport tourism: sport tourism as a novel tourism phenomenon (Kurtzman & Zauhar, 1997), the definition of sport tourism (Gammon & Robinson 1997) and the sport event as a service and demand model (Getz, 1998). As a result of Hall and Page's (1999) extensive literature analysis, researchers have explored the relationship between recreation and tourism in the following contexts:

• Knowledge – content of culture, science, tourism training;
• Action – examination of behaviour, test practice, tourism training and
• Culture – society, research community, tourism students.

Dreyer (2002) analysed the effects of major sporting events in various sociological, health, ecological and geographical aspects involving different disciplines. In these model descriptions, the authors stated that sport tourism played a significant role in the discovery of new areas, the exploration of environmental impacts and regional spatial development.

Managing Elements of Sport Tourism

Many tourism providers believe that it is enough to construct sport facilities around their lodging services and later on they wonder why tourists don't use them. We think much more should be offered, which is connected to managing the sport-related aspects.

Based on the science and practice, we have set up a sport tourism management model to exemplify the elements of tourism, which can assist researchers and practitioners in their management work (Figure 1.2). The items in the illustration are closely related to each other, so we do not consider it necessary to rank them.

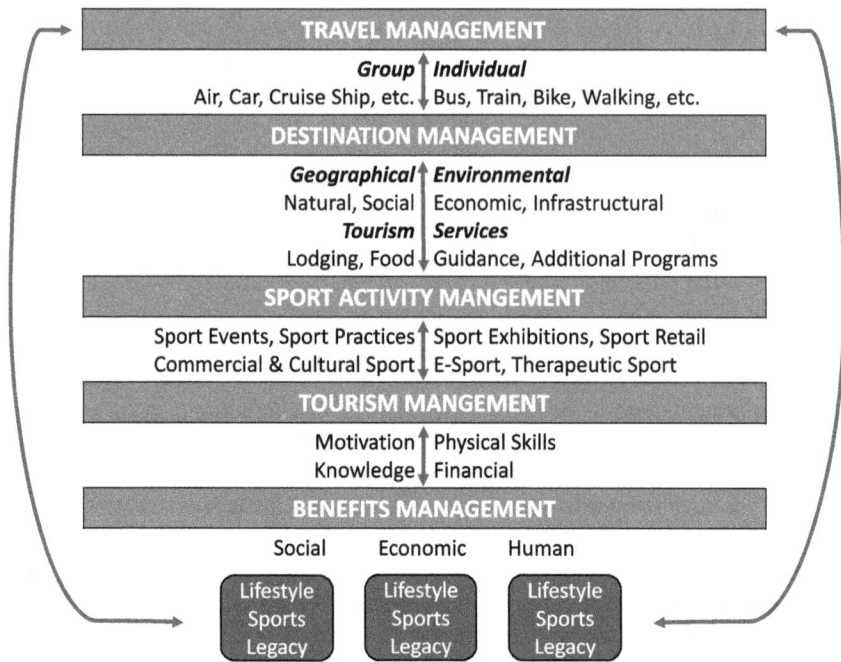

Figure 1.2 Sport Tourism Management Model.

Unlike economists and geography scientists, sports scientists state that the focus of sport tourism is based on active organizers (animators) and participants (tourists) who are physical and motivated to the environment. The process of their joint activities determines the success of the sports tourism programme, which is based on the following key elements:

Travel Management

Tourism – also reflected in its name (tour – English word) – is about travel, in which mode of travel plays a key role. According to the Travel Industry Association of America (2004), in 2003, people in the United States spent $554 billion on local and international travel, representing 7.2 million jobs. Of this, foreigners spent US$80.2 billion on their US travel and American's spend US$77.6 billion abroad.

Also in sport tourism, travel is one of the key elements and can be both passive and active. Passive travel means the transport of passengers by public transportation system (car, train, bus, air), which has significantly increased.

This has opened up new tourist areas to the world. Due to the fall in ticket prices, the participation of young people in tourism has increased, leading to a significant increase in the number of people interested in sports tourism.

In sports science, issues related to passive travel arise where the fatigue of travel and the associated time difference affect the level of athletic ability.

Active travelling is when tourists can reach their destination by involving their physical abilities (on foot, by bike, by boat, etc.). Not only do tourists reach their destination, but they also get to know the environment during their travels, experience movement and have the opportunity for intensive socialization.

Destination Management

In sport tourism, one of the most influential elements is the geographical environment of destinations, which can be different from one's home. These can be natural, social, economic and infrastructural factors (Kurtzman & Zauhar, 1997), which influence the comfort of a stay for tourists. It may also increase the effects, changing the intensity of sports activities (Hamilton & Banhidi, 2004).

Sports physiological research results provide rich information on the effects of different natural resources as part of the natural environment. For example, the increase in load due to elevation changes can be accurately described (Hamilton, 2000). Tourists arriving in the mountains are also known to experience physiological effects of altitude atmospheric changes. As altitude increases, the air pressure and the oxygen content of the air decreases, making it difficult for the body to absorb oxygen. At 3,650 m, the air pressure is approximately 40% lower than sea level air pressure; this also means less inhaled oxygen. At 2,400 m, oxygen uptake is 12% lower, 3,100 m is 20% lower and 4,000 m is 27% lower (Pavia & Kennedy, 2005; Powers & Howley, 1997).

A similar increase in exercise occurs with a difference in physical activity in the climatic environment. The temperature of the human body responds to changes in the external temperature at variable rates. During sport, the human body produces a significant amount of heat. At a load of 20–30%, 70–80% of the energy produced creates heat and can be multiplied by increasing the intensity. For example, at 25°C cycling at maximum intensity for one minute, the skin temperature rises to 28°C, which, even after stopping the load, increases linearly above 30°C for at least 10 minutes, 20% higher than the original ambient temperature (Figure 1.3).

To compensate for this, the body responds with swelling, ventilation and radiation. This is slower in cold and quicker in warm environments, so knowing the sporting environment requires careful attention to whether or not the outdoor environment is very warm, in addition to humid. For example, near the Mediterranean coast, heat dissipation will be significantly more difficult. In such cases, the conditions of the sport are significantly burdened by the shifting of the body's heat balance. Exercising at high ambient temperatures can have a number of dangerous effects, causing the body to overheat faster (47°C), which can lead to hyperthermia, sunstroke, heat cramps, heat stroke, even circulatory failure and death (Howley & Franks, 1992).

The effect of wind, which is an element of the weather, has a variable effect on tourist activities. For some, it is enjoyable, and for others, it causes a lot of inconvenience. However, for sports tourism, knowledge of wind effects is important because:

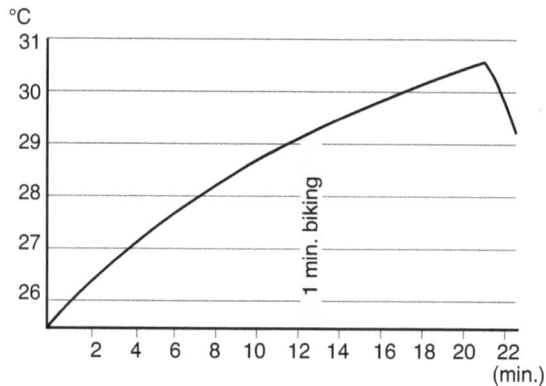

Figure 1.3 Changes of Skin Temperature During One Minute Biking Test.
Source: Banhidi (2004b).

- Security forecast – a trigger for weather changes,
- Load increase – increase of air resistance (e.g., cycling, inline skating),
- Change in sensation of heat – significant stress in cold environment and
- Optimal conditions for sailing sports – propulsion (e.g., sailing, gliding, parachuting).

Gloomy, rainy weather can be a nuisance for many tourists, but it can also be positive for sports tourism. In fact, sport science studies have shown that it is possible to achieve better physical performance in humid, drizzling rain (Trapasso & Cooper, 1990; Zhang et al., 1991), which could even be incorporated as a tourist attraction in sport tourism.

The influential role of the social environment in sport tourism is significant in terms of different regulations, traditions and customs. Researchers focus on understanding the socio-cultural background of sports activities, in which participants are in constant interaction with the social environment and tourism providers (Kosiewicz, 2006). Popular sports play an important role in the selection of tourist destinations and are considered to be a valuable aspect of sport tourism attraction. Tourists are looking for the culture of tennis at Wimbledon, for winter sports in the Swiss mountains or for adventure tours in Africa. There are other expectations for a tourist in an urbanized or rural setting. While the former prefers the more advanced sports facilities, the latter prefers the opportunity offered by the natural environment.

The role of infrastructure in sport tourism is appreciated by professionals in many ways, and we believe it deserves much more attention. It can be characterized bilaterally. On one hand, the main question is how sports facilities, which were built to host major sport events, can be integrated into the circulation of tourism, and on the other hand, how sports facilities built around tourism providers meet the expectations of sport travellers.

In ancient times, stadiums were built to accommodate as many guests as possible coming from further distances. The Panathenaic Stadium in Athens is still a major tourist attraction, with various tourist holidays. In Barcelona, the Olympic Stadium, which was designed in 1929, is open to tourists year-round and today some hotels advertise using the stadium's image (e.g., Silken Concordia, Fira Palace, Catalonia Plaza). Hotels, gymnasiums, tennis courts, swimming pools and riding stables in the area are proving that the accommodation can be sold better in its presence. Many authors point to this growing phenomenon, but accurate research data is only sporadically available.

Sport Activities Management

In the centre of the sports tourism model and from a sports science point of view, we consider sport activity itself. In sport tourism management, sport activities are important from two sides. When it comes to passive sports tourism, the organizers need only be aware of the social aspects of sport (history, rules and conditions of sports competition). However, in the case of active sports tourism, you should be aware of the motivational elements of sports activities, as well as the physiological and pedagogical aspects of movement in relation to differences from competitive sports (Table 1.2).

Knowledge of the intensity of sporting activities in tourism is essential as it can significantly affect the athlete's well-being. When the intensity reaches the training zone during a sporting activity, it can eventually lead to unpleasant physical effects, which can even lead to activity termination (Hoeger & Hoeger, 2015).

At tourism destinations, many misjudge tourists' sport motivation. Uncertainty is usually caused by the fact that developers often rely on popular sports rather than offering big variety of sports using their local resources. In the Alps, most sports tourists are looking for mountain activities; at the seaside, they are looking for water sport activities.

Benefits of Sport Tourism

In international literature, much is written on the economic and social impacts of certain types of major sports tourism event such as the Olympic Games and soccer

Table 1.2 Characteristics of Sport Activities in Competition and in Tourism

Sport in competitive sport	Sport practice in tourism
Winning is primary	No need to compete
The need to increase performance	There is no compulsion to improve performance
Reaching the maximal capacity	Health load
Standard condition	Changing conditions
Well-known environmental factors	Unknown environmental conditions
No mistake is allowed	Mistakes are allowed
Designed and directed by others	The need for self-organization is common
Health is secondary	Health is primary

World Cup, *Tour de France*. Usually, the vast majority of residents are aware of the event with many planning to watch the race or participate in related activities. Furthermore, despite the potential for various negative impacts, there was overwhelming support for the decision to host the event, suggesting a very successful promotional campaign by the city council (Bull & Lovell, 2007). Especially in metropolitan areas (such as Chicago, London, New York, Paris, Tokyo), the benefits come from ample labour supplies and sophisticated transportation (Daniels, 2007).

We believe that many parties can benefit from sport tourism development. It works best if the providers involve the values of sport in their services, such as social, economic infrastructural legacy or human benefits. According to Hritz and Ross (2010), social and economic benefits are strong predictors of support for further sport tourism development. This reveals a strong identification with the advantages of sport tourism in their city, such as an increased cultural identity and social interaction opportunities. Other studies underline the main effects of tourism (Hall & Page, 1999), which also appear in the sport tourism sector (Table 1.3):

In active sport tourism, the physical benefits are the most unique, which are generated by the intensity of physical activities. Those activities are the most effective if they are planned and led by professionals. According to sport physiologists, organized training sessions and competitions develop athlete's psychomotor abilities, measured by many sport scientists (Cioni & Sgandurra, 2013; Foss & Keteyian, 1998). Therefore, tourism needs to examine the parameters of the load of sport tourists to avoid uncomfortable or negative benefits of exercise.

The results of our former research have shown that during cycling, skiing or kayaking, tourists choose a comfortable intensity (50% load) and a pleasant posture, to enjoy the movement, the feeling of freshness and circulation (Banhidi, 2002, 2003). It is also known that sport tourists don't have the same physical conditions

Table 1.3 Effects of Sport Tourism

	Natural environmental impact	*Social impact*	*Economic impact*	*Human impact*
Positive	Increased consciousness Protection of wildlife Better knowledge on nature Take advantage of terrain	More contacts Increased knowledge Development of sport infrastructure More varying occupations	Job opportunities Higher income of host communities	Higher physical abilities Better skills More knowledge on sports Better health level
Negative	Erosion Littering Pollution Exploring to untouched areas	Alienation Criminality "Disneyfiction" of culture	More seasonal jobs Rise in prices Increase dependence Costs of development	Risky situations Accidents It distracts from other attractions

Source: Modified from Hall and Page (1999) and Pettersson (2004).

or former experiences, so the benefits occur differently. For example, an advanced (non-racing) skier needs less energy to ski downhill on a slope than a beginner who needs more stops to rest. The unexperienced sport tourist can experience more injury. Many tourists are not aware of risks of injury, so preparations and guidance are needed in the sector.

References

Ardern, C. L., Glasgow, P., Schneiders, A., Witvrouw, E., Clarsen, B., Cools, A., … Mutch, S. A. (2016). 2016 consensus statement on return to sport from the First World Congress in sports physical therapy, Bern. *British Journal of Sports Medicine, 50*(14), 853–864.

Bang, H., & Eaton, L. (2006). *Volunteer motivation and commitment at a sport-tourism event.* University of Minnesota. Retrieved March 12, 2018, from https://bit.ly/2DjucjC

Banhidi, M. (2002). *Testing bike trails and bikers for bike tourism development.* NYME AK, Győr: Apáczai Napok Tanulmánykötet.

Banhidi, M. (Ed.). (2003). *Water and beach tourism. [A vízi és vizekmenti turizmus alapjai.]* Budapest: Faculty of Economy and Tourism.

Banhidi, M. (2004a). Promoting wellness: European perspective. In: M. K. Chin, L. Hensely & P. S. H. Cote Chen (Eds.), *Global perspectives in the integration of physical activity, sport, dance and exercise science in physical education: From theory to practice* (pp. 133–144). Dept. of Physical Education and Sports Science, The Hong Kong Institute of Education.

Banhidi, M. (2004b). Sport and cartography. *Journal of Coimbra Network on Exercise Sciences, 1*, 57–63.

Brämer, R. (2002). Megatrend Wandern – Problem oder Chance? Sport und Tourismus Dokumentation des 10. *Symposiums zur nachhaltigen Entwicklung des Sports vom 28–29 November in Bodenheim/Rhein.*

Brown, R. M. (1935). The business of recreation. *Geographical Review, 25*, 467–475.

Bull, C., & Lovell, J. (2007). The impact of hosting major sporting events on local residents: An analysis of the views and perceptions of Canterbury residents in relation to the Tour de France 2007. *Journal of Sport & Tourism, 12*(3–4), 229–248.

Buning, R. J., Cole, Z., & Lamont, M. (2019). A case study of the US mountain bike tourism market. *Journal of Vacation Marketing, 25*(4), 515–527.

Carlson, A. S. (1938). Recreation industry of New Hampshire. *Economic Geography, 11*, 977–988.

Cioni, G., & Sgandurra, G. (2013). Normal psychomotor development. *Handbook of Clinical Neurology, 111*, 3–15.

Crisler, R. M., & Hunt, M. S. (1952). Recreation in Missouri. *Journal of Geography, 51*, 30–39.

Daniels, M. J. (2007). Central place theory and sport tourism impacts. *Annals of Tourism Research, 34*(2), 332–347.

Deasy, G. F., & Griess, P. R. (1966). Impact of a tourist facility on its hinterland. *Annals of the Association of American Geographers, 56*, 290–306.

DiFrancisco-Donoghue, J., & Balentine, J. R. (2018). Collegiate eSport: Where do we fit in? *Current Sports Medicine Reports, 17*(4), 117–118.

Dilek, S. E. (2019). E-Sport events within tourism paradigm: A conceptual discussion. *Uluslararası Güncel Turizm Araştırmaları Dergisi, 3*(1), 12–22.

Dreyer, A. (2002). *Sport und Tourismus. Wirtschaftliche, soziologische und gesundheitliche Aspekte des Sport-Tourismus.* Wiesbaden: Universitätsverlag.

Eventseye. (2022). *Sport – Animals trade shows 2022–2023*. Retrieved from https://www.eventseye.com/fairs/st1_trade-shows_sports_8.html

Fletcher, M. (2016). *How global sporting events score economic goals*. Retrieved November 23, 2019, from https://bit.ly/37MXs0q

Foss, M. L., & Keteyian, S. J. (1998). *Fox' physiological basis for exercise and sport*. Boston, MA: WCB McGraw-Hill.

Funk, D. C., & Bruun, T. J. (2007). The role of socio-psychological and culture-education motives in marketing international sport tourism: A cross-cultural perspective. *Tourism Management, 28*(3), 806–819.

Gammon, S., & Robinson, T. (1997). Sport and tourism: A conceptual framework. *Journal of Sport Tourism, 4*, 8–15.

Geffroy, V. (2017). 'Playing with space': A conceptual basis for investigating active sport tourism practices. *Journal of Sport & Tourism, 21*(2), 95–113.

Getz, D. (1998). Trends, strategies, and issues in sport-event tourism. *Sport Marketing Quarterly, 7*, 10–13.

Getz, D., & McConnell, A. (2011). Serious sport tourism and event travel careers. *Journal of Sport Management, 25*(4), 326–338.

Gibson, H. J. (1998). Sport tourism: A critical analysis of research. *Sport Management Review, 1*(1), 45–76.

Green, B. C., & Chalip, L. (1998). Sport tourism as the celebration of subculture. *Annals of Tourism Research, 25*(2), 275–291.

Hall, C. M., & Page, S. J. (1999). *The geography of tourism and recreation*. London, New York: Routledge.

Hamari, J., & Sjöblom, M. (2017). What is eSports and why do people watch it? *Internet Research, 27*(2), 211–232.

Hamilton, B. (2000). East African running dominance: What is behind it? *British Journal of Sports Medicine, 34*(5), 391–394.

Hamilton, N., & Banhidi, M. (2004). *Multi-dimensional mapping for popular sports. II*. International Conference for Physical Educators (ICPE 2004) 7–10 July, p. 75.

Harris, S. R. (2012). "We're all in the same boat": A review of the benefits of Dragon Boat racing for women living with breast cancer. *Evidence-Based Complementary and Alternative Medicine, 2012*.

Higham, J. (2018). *Sport tourism development*. Toronto: Channel View Publications..

Higham, J. E. S., Kearsley, G. W., & Kliskey, A. D. (2006). *Multiple wilderness recreation management: Sustaining wilderness values- maximising wilderness experiences*. Electronic publication. University of Otago.

Hoeger, W. W., & Hoeger, S. A. (2015). *Principles and labs for fitness and wellness*. Wadsworth: Cengage Learning..

Howley, E. T., & Franks, B. D. (1992). Health fitness. *Instructor's Handbook*. Champaign, IL: Human Kinetics.

Hritz, N., & Ross, C. (2010). The perceived impacts of sport tourism: An urban host community perspective. *Journal of Sport Management, 24*(2), 119–138.

Hutchins, B. (2008). Signs of meta-change in second modernity: The growth of e-sport and the World Cyber Games. *New Media & Society, 10*(6), 851–869.

Jisha, J., Mozen, D., & Talbot, K. (2004). Organizing a girls sport expo for an elementary or middle school. *Strategies, 17*(6), 11–12.

Jones, S. B. (1933). Mining and tourist towns in the Canadian Rockies. *Economic Geography, 9*(4), 368–378.

Kay, T., & Jackson, G. (1991). Leisure despite constraint: The impact of leisure constraints on leisure participation. *Journal Leisure Research, 23*, 301–313.

Kosiewicz, J. (2006). Tourism from social perspective. In J. Kosiewicz (Ed.), *Environmental differentations of tourism* (pp. 7–8). Warsawa: Legionowo, Wydawnictwo i Ksiegarnie.

Kurtzman, J., & Zauhar, J. (1997). A wave in time–the sports tourism phenomena. *Journal of Sport Tourism, 4*(2), 7–24.

Kurtzman, J., & Zauhar, J. (2005). Sports tourism consumer motivation. *Journal of Sport & Tourism, 10*, 21–31.

Lamont, M., & McKay, J. (2012). Intimations of postmodernity in sports tourism at the Tour de France. *Journal of Sport & Tourism, 17*(4), 313–331.

Lang Research Group. (2004). *Interest of professional sports as a spectator.* Ministry of Tourism. Retrieved September 15, 2007, from https://tvv-journal.upol.cz/artkey/tvv-200801-0011_SPORTOVY_TURIZMUS_NA_SLOVENSKU_A_POROVNANIE_JEHO_ZVYKLOSTI.php

Neirotti, L. D., Bosetti, H. A., & Teed, K. C. (2001). Motivation to attend the 1996 Summer Olympic Games. *Journal of Travel Research; Boulder, 18*, 2–8.

Panja, T. (2014). *Rio jilts World Cup as $11 billion bill sours Brazil.* Bloomberg Business. Retrieved April 10, 2024 from https://www.bloomberg.com/news/articles/2014-05-28/rio-jilts-world-cup-as-11-billion-bill-sours-brazil?embedded-checkout=true

Pavia, W., & Kennedy, D. (2005). *Highest peak continues to hold fatal attraction.* Retrieved December 25, 2005, from https://bit.ly/2XOH4HT

Pettersson, R. (2004). *Sami tourism in northern Sweden: Supply, demand and interaction.* [Doctoral dissertation] Umeå universitet.

Pigeassou, C. (2004). Contribution to the definition of sport tourism. *Journal of Sport & Tourism, 9*(3), 287–289.

Powers K., & Howley E. T. (1997). Exercise physiology by scott. Brown & Benchmark New York: McGraw-Hill Education.

Röthig, P. (Ed.). (1992). *Sportwissenschaftliches Lexikon.* Schorndorf: Hofmann.

Saayman, M. (Ed.). (2012). *An introduction to sport tourism and event management.* Bloemfountain: Sunpress.

Selke, A. C. (1936). Geographic aspects of the German tourist trade. *Economic Geography, 12*, 206–216.

Simpson, A. (2005). *The London Olympic Games 2012. Tourism marketing intelligence.* Washington, DC 18th Nov. Retrieved November 7, 2016, from http://www.comlinks.com/tourism/ti051118.htm

South African Government Information. (2007). *Sports tourism project.* Retrieved March 18, 2010, from https://bit.ly/2Oln4K8

Statista. (2025). *eSports market revenue worldwide from 2020 to 2025.* Retrieved April 10, 2024, from https://www.statista.com/statistics/490522/global-esports-market-revenue/

Trapasso, L. M., & Cooper, J. D. (1990). Record performance at the Boston Marathon: Biometeorological conditions. *International Journal of Biometeorology, 33*(4), 233–237.

Turco, D. M., Riley, R., & Swart, K. (2002). *Sport tourism.* Morgantown: Fitness Information Technology, Inc.

United States Department of Agriculture. (2003). *National forest visitor use monitoring program. National project results January 2000 through September 2003,* USDA Forest Service, Washington, DC.

Wolfe, R. J. (1967). Recreational travel: The new migration. *Geographical Bulletin, 9*, 73–79.

Woods, B. (2004). *Applying sport psychology to sport.* London: Hodder & Stoughton.

Zachazewski, J. E., & Magee, D. J. (Eds.). (2012). *Handbook of sports medicine and science: Sports therapy: Organization and operations* (Vol. 19). Hoboken, NJ: John Wiley & Sons.

Zhang, S., Meng, G., Wang, Y., & Li, J. (1991). Study of the relationships between weather conditions and the marathon race, and of meteorotropic effects on distance runners. *International Journal of Biometeorology, 36,* 63–68.

2 Theory of Sport Tourism

Findings of the Past Decades

Miklos Banhidi

Introduction

Sport tourism theory is a fast-growing discipline that requires constant detailed analysis to understand its foundation and development. It is a young tourism sector, where social and economic impacts are becoming increasingly dominant in destination development. The founding of sport tourism theory was the focus of many researchers from different disciplines, such as geographers on sport tourism environment, economist on financial benefits, sociologists on social influences, and psychologists on tourism motivations and lifestyle habits. There are only a few publications by sport scientists on how the sport industry fits into tourism services, or how much sport activities differ from home activities. So, in this chapter, we have considered analyzing publications from that perspective too.

In our recent study, we conducted a comprehensive bibliometric and content analysis by reviewing 811 publications (62 articles per year) collected by web-based scientific search program between 2005 and 2017 to discover researchers' focus and their reported results. Based on the content analysis, 11 categories were formulated to observe the frequency of the yearly changes in research topics. Content analysis is a fast-growing scientific method that allows researchers to examine a discipline by analytical methods, giving direction to future research (Edginton et al., 2014; Neuendorf, 2002). The aim of our study was to analyze the field of sports tourism so that we can better understand the processes in the area of social, economic, and human impacts. Our results will be beneficial to sports scientists.

Based on the rankings, the majority of researchers wrote about the motivation and behavior of sports tourists, as well as the social effects of sport tourism. There are very few publications that can be found on active sports tourism, which indicates that few sport scientists have been linked to that research area. It is difficult to find studies on the human impact of sports tourists, so we believe that, in the future, researchers should pay more attention to the benefits of sport in the field of tourism.

We are convinced that sports tourism has practically always existed in practice, with the spread of sport competitions, and with many sports facilities built in holiday destinations. From the ancient Olympic Games to modern sport competitions (e.g., sailing competitions in 1661, racing events in 1843, bicycle competitions in 1863), visitation has emerged from a variety of athletes and spectators alike. This

DOI: 10.4324/9781003476658-3

trend has been further intensified as a result of globalization, expanding sporting competitions and travel opportunities. The main goal of traveling was to be a part of the sport industry and to visit sport events, which increased the importance of on-site tourist services.

In one of our previous studies, we analyzed components of the sport tourism sector considering sports scientific perspectives (Banhidi, 2007). It has become clear to us that the definition of sport tourism is seen differently by sport scientists as opposed to economists and sociologists. Sports scientists consider sport as a cultural activity that requires psychomotor abilities, which are influenced by biologic development and training processes or external factors such as socio-economic elements and technological changes. In sport tourism research, we consider sport activities as tourism motivation and an excellent opportunity to fulfill most traveler expectations. Through practicing sport tourism, tourists can rest more easily, experience challenges, and learn more about nutrition. These benefits can be even maximized through some supplemental tourism activities that interplay not only within the different categories of sport tourism attractions but also across non-sport tourism attractions (Ito & Higham, 2020).

It is almost commonplace that sporting events appear in tourism destinations as alternatives to local entertainment, which supply additional revenues to service providers. Sport tourism experts predict exponential opportunities in this sector since they have discovered links to other tourism sectors. Among these links is the fact that tourists are getting to know the destination much better by engaging in sport tourism activities. Visitors experience and understand the culture and function of the host locations; therefore, it is understandable that more and more sports facilities are being built in tourism centers to increase destination competitiveness.

Former Publications on Sport Tourism Development

Research projects on the development of sport tourism increased in the 1990s with the aim of finding answers to the following questions (Ross, 2001):

- How can the forms of sport tourism be interpreted as sport, tourism, active, and nostalgic sports tourism?
- Why has the sport tourism sector become so popular?
- What are the direct and indirect influencing factors?
- What are the characteristics of sports tourists?
- How can sport events be organized successfully and how can they be sponsored?

These issues were aimed to understand the growth of this booming sector, but at that time, there were still few synthetic studies on the relationship between the two mega sectors: sport and tourism.

In our opinion, the definition of sport tourism is crucial for understanding the science of both sectors. When discussing both sport and tourism, several common values appear, such as happiness, enjoyment, satisfaction, relaxation, socialization, human and economic development.

Since the 1920s, sport scientists were seeking to find better research methods for analyzing athletes and their skills (Kent, 1994), primarily to achieve peak performance. This means that there was always a need to involve researchers to educate top athletes (Sydney Sports and Exercise Physiology, 2010), requiring specialists from different scientific disciplines. During the development of sports science, the research addressed any knowledge of the sports sector where socio-economic interests appeared. These included, for example, amateur athletes, their organizations, and their different lifestyles (Slack & Hinings, 1987). Among others, sport psychologists have not only measured top athletes, but also people practicing recreational sports, underlining the importance of health benefits (De Pauw et al., 2013).

For scientific understanding of tourism, researchers use different theoretical approaches regarding the environment, economics, and technology. This clearly reflects the entry of different disciplines, such as geography and economics. According to some researchers, a holistic approach is needed, involving several elements such as the tourist, geographic factors, the destination, and the tourism industry (Leiper, 1979). In another approach, the theoretical knowledge of tourism is ranked among the social sciences (Wang, 1998). Accounts of various science disciplines appear and, therefore, tourism can be labeled an interdisciplinary field of research. Perhaps, as a result of this approach, the *International Journal of Tourism Sciences* was founded in 2000, anticipating research studies from a number of tourism disciplines.

The general scientific interest to better understand sport tourism has risen since 1990, mainly due to the dynamic development of the sector (Hinch & Higham, 2001). Since 1993, under the auspices of a private Canadian organization, the Sport Tourism International Council has provided an online source of information (Gibson, 2003). The results have also been published in other professional journals and books. These include *Tourism Recreation Research* (1997), *Journal of Vacation Marketing* (1998), *Vision in Leisure and Business* (1999), etc.

The importance of this subject area is supported by the fact that the number of tourists seeking sport opportunities has increased significantly. These tourists have also been willing to spend more on sports services, which has already economically benefited other tourism sectors. For example, in Australia, the tourism industry has already accounted for 5% of sports tourism products with $3 million spending. In order to enhance this, a national sport tourism strategy was developed in 2000 (Foo, 2000).

More significant shares have been registered in the United Kingdom, where 20% of travelers consider sport as a primary motivation for travel.

As a result of the theoretical development of sport tourism, in 2001, the World Tourism Organization held a large conference together with the International Olympic Committee. The conference held presentations and discussions that resulted in a general recognition of the close relationship between sports and tourism when more researchers were involved in analyzing the sector.

Besides traveling, the definition of sport includes gaining experiences (Aho, 2001), to learn more about sports culture (Pigeassou, 2004), in which understanding plays a leading role. This is particularly characteristic of sports tourism products where human benefits are greatly enhanced by sight and movement. At the beginning, research focused on the abundant knowledge of the developmental effects of sporting events. This research only indirectly affected human experience and highlighted the advantages of the sport profession. From a scientific point of view, however, there were remarkable findings that revealed a widening in the target groups of interest as a result of the development of local sports culture in sporting events and local sports services.

Some studies on these target groups have reported that most active travelers prefer traveling alone. The studies also revealed that active travelers are ready to travel even longer distances to find better destinations for their favorite sport, to which they will return more than once. The largest percentage of event spectators are those who are involved in sporting events every day (Ross, 2001), such as reading sports news and watching sports reports. At the same time, there is an increase in the number of travelers who are looking for pleasure with extreme activities and taking the risk (Elsrud, 2001).

The significance of sport tourism was often interpreted by scientists in conjunction with other scientific disciplines. For example, in a sociological approach to sports, sport tourism was interpreted in terms of spatial and temporal dimensions (Hinch & Higham, 2001). In their view, the interpretation of spatiality is justified by the fact that the experience of sport is taking place in an environment alternative to home, with different natural, social, and economic characteristics. It is also necessary to characterize the temporality of sports experience in order to discover the duration of the travels. For example, the number of guest nights spent or the number of services used. However, there is currently no research examining how much physical or mental load is involved in longer periods of travel, or how much more calorie intake or after-load relaxation is needed.

The exact definition of the sector is complicated by the fact that sport tourism, as a part of the sport industry, rarely appears as an independent product. Sport tourism is most often linked with other tourism sectors as a complementary activity (Turco et al., 2002).

Previously, the World Tourism Organization did not consider sport tourism as a separate sector, despite the fact that ski resorts and mountain walking trails were already dominant in destinations of mass tourism. In the last few decades, several studies address a new interpretation of sport tourism. This is due to the fact that both the sports industry and the land use designated for sports have been expanding rapidly (Hinch & Higham, 2010). In our opinion, since 2005, the number of sport tourism studies have increased because of the emergence of more sports-related industrial sectors within tourism (e.g., training, competition systems, and sporting events). Tourism providers have recognized that tourism experiences can be improved by offering special services, including unique travel opportunities, educated guides, supplementary programs, or uncommon sites and infrastructure.

The model above (Figure 2.1) is an optional list of sport tourism products, which introduces the tourism motivations, where the demands of environmental stimuli and new activities appear more intensively. Within this sector, there is an increasing demand for physical, mental, commercial, and therapeutic recreation.

Figure 2.1 Sport Tourism Motivation.

There is also an increasing demand for tourism products based on a wide variety of activities, which has resulted in significant growth of the services tourism sector (Bonzak, 2013). One of the priorities of travelers is to live a healthy lifestyle while traveling, possibly even healthier than at home. More and more people recognize that practicing sports can be a valuable tool to achieve a wholesome lifestyle. There are also increasing expectations by sport travelers to experience challenging and sometimes risky activities. This niche form of tourism is called adventure tourism, and despite targeting a small demographic, the industry's growth is already well known (Kane & Tucker, 2004).

Specialists working in the sport sector have also recognized the opportunities offered by tourism. The turnover of sport stores proves that in the tourism season, a large amount of sports equipment can be sold (Starr, 2000). For example, many people buy sports equipment before traveling and if they don't find what they want at home, they are ready to buy it at their destinations (Banhidi, 2011). Some of the modern stadiums provide full services to the guests traveling longer distances.

At many sports developments, it is even believed that their facilities and programs are also suitable for hosting tourists. It is also no coincidence that numerous training centers were established around the world. For example, training centers were developed for long distance runners in Kenya, for European skiers in New Zealand, and for water sport athletes in South Africa.

However, professionals point out that the development of sport tourism is constantly influenced by socio-economic and ecological changes, so the significance of the sector must be frequently re-interpreted (Hinch & Higham, 2010). This is linked to climate change, as winter sports must move to the higher altitude areas and aquatic and terrestrial sports become affected by extreme weather changes.

The bibliometric data of the published sports tourism studies are shown in Figure 2.2. As a result of the content development, we identified 11 categories according to the main focus of the researchers' investigations. Here, we found overlaps in several cases, so there were studies that could be included in more than one category.

In the last decades (N-811) we found four to five publications each year, which concerned the theoretical definition, point of attachment, and research methods of sport tourism. This demonstrates the need for continuously updating the theoretical approach of the sport tourism sector.

Most of the studies describe the characterization, motivation, experiences, and travel patterns of sports tourism (16.4%). Many of them noted that the sports tourism industry is now offering opportunities for anyone – people of all ages and genders – who would like to engage in their passions or seek out challenges and experiences. Interestingly, only a few people have written about the importance of learning (2.9%) and health awareness and quality of life (2.2%) within the participation of sports tourism.

Many studies (13.9%) report the role of social issues in sport tourism and mention that sporting events have a significant effect on society (9.9%), and that influences the development and operation of host communities. Many of the scholars emphasize the fact that sporting events leave a number of legacies, such as the growth of sports infrastructure, development of local sports culture, and increased criminal statistics for events. We found an interesting term in one of the studies:

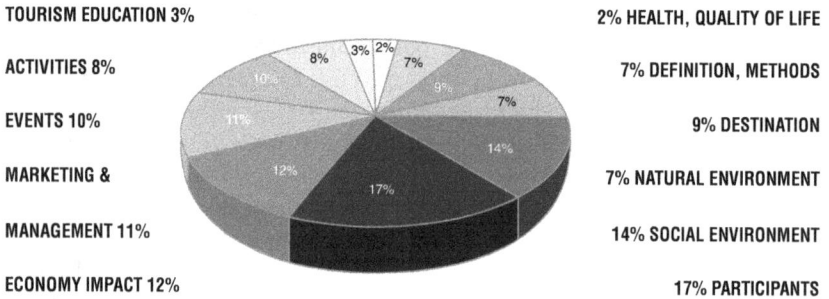

TOURISM EDUCATION 3%

ACTIVITIES 8%

EVENTS 10%

MARKETING &

MANAGEMENT 11%

ECONOMY IMPACT 12%

2% HEALTH, QUALITY OF LIFE

7% DEFINITION, METHODS

9% DESTINATION

7% NATURAL ENVIRONMENT

14% SOCIAL ENVIRONMENT

17% PARTICIPANTS

Figure 2.2 Distribution of International Publications by Sport Tourism Topics (N-811).

"lifestyle migration". This term characterizes people who travel to sports tourism destinations where they can play their activity of choice at a higher level.

Many researchers wrote about regional sport tourism development opportunities (9.3%), as well as the relationship between sports and ecotourism (7%) and the importance of the natural environment for sports tourism products. At the same time, attention is also drawn to the consequences of climate change.

A significant number of studies are based on an economic approach that assumes sporting events and enterprises dominantly influence the local economy (12.2%). To enhance this, many have written about the outstanding role of management and marketing (10.9%).

During the period under review, studies have examined areas of sports tourism in a wide variety of approaches, often quoting earlier publications to ensure results are clearly built on each other. In the last two years, most publications focus on the characterization of sports tourism destinations and sports tourism as a whole, reflecting changes in the interest of researchers but not suggesting material differences from previous years.

Research Results on Sport Events Tourism

There is no doubt about the attractiveness of sports competitions in the realm of tourism. Studies in this regard were aimed at the motives, habits, and information on the use of the versatile visits to local viewers and tourists. However, many scholars have also measured the difference of traveling habits between sport tourists and other tourists. According to the results, for visiting, sightseeing, and the quality of sports facilities, sports enthusiast visitors are much more demanding than the local spectators (Chen & Funk, 2010). Research data has also revealed that sport tourists are more interested in the details of the sport sector, using more services and spending more money.

According to a study of the motivations of sports tourism in the USA, there are the highest number of spectators who take a journey further afield. According to the results of factor analysis research of foreign students, the main motives of their trip were the travel experience, the interest in professional sports, the spectacle of sporting events, and cheering for their favorite players (Yu, 2010).

Researchers often asked questions about whether a sporting event could be a tourist event on its own. The simplest answer would be yes, but it depends on whether the competition attracts tourists or not and whether or not they come from another location. Behind the debate, there is also the impression that the spectator has the motivation to travel even further to visit a sports competition. It is also important to note whether there is a high-quality alternative program such as a special service at the match. In Japan, the importance of sports spectators is categorized into how far the spectators have come from (Banhidi, 2011), so we could classify their tourism significance accordingly.

Another explanation might be that most tourists appear at the most popular competitions (Figure 2.3). These competitions could be the National Football League (NFL), the Bundesliga, or the English Premier League (EPL), which attract

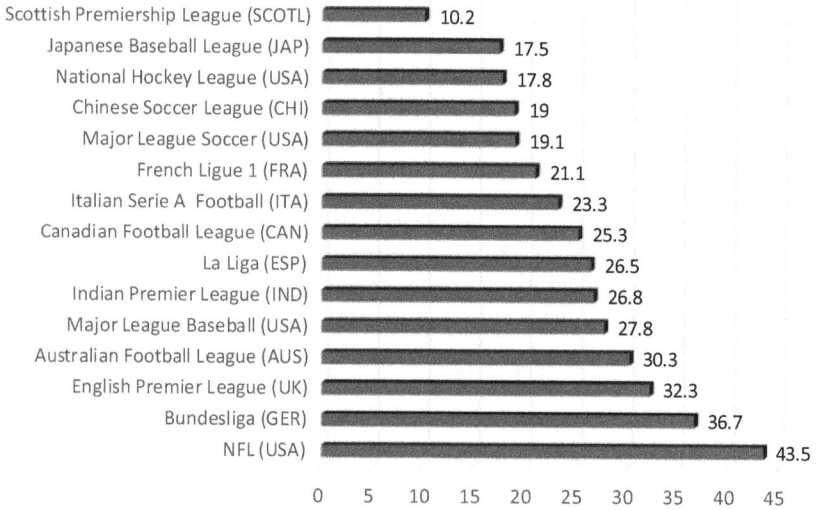

Figure 2.3 The Most Visited Leagues in the World in 2014/2015.
Source: Gaines (2015).

millions of spectators every year. Although there is not always accurate data on the number of tourists attending the events, estimates clearly show that at least two-fifths of all US residents are sports travelers, and 84% of them are frequent spectators. Fifty-eight percent of sport tourists gladly return later with their family to the venue destinations (Wittmann, 2017).

Professionals in sport economy have dealt with a number of studies regarding how the integration of the sports sector in tourism has resulted in profits (Ritchie & Adair, 2004). Since ancient times, the need for spectacles is still very strong among sport spectators (Kyle, 2015). People are ready to travel to those destinations where they can experience a unique environment, and they can challenge their physical limits.

Many sports tourism researchers have been looking at the event market because it is a clear example of the benefits of tourism services. The upsurge of tourism is most noticeable at major events, which are mostly dominated by the spectators and not as much by the location itself. The importance of passive sports tourism is supported by data in the early 2000s that states, 75% of British and 44% of Australian adults visited sport events. In the United States, at the world's most visited sports events, the average number of viewers was about 72 million, which is 35% of all adult citizens (Sporting Intelligence, 2015). Also, many European Union citizens enjoy visiting sporting events, but this differs by country (Figure 2.4). The highest percentage of citizens who watched a sporting event at least once a year are from the Netherlands, Norway, Sweden, Switzerland, Finland, Iceland, Ireland, and the Czech Republic. These values clearly demonstrate the importance of the mobility and economic influence of sporting events.

Percentages of Sport Events Visitors in the EU
(2016, %)

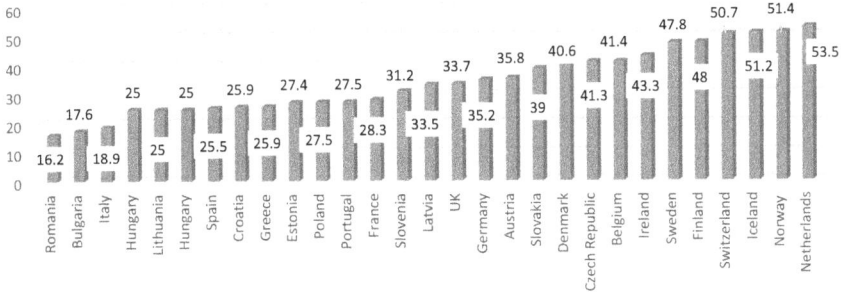

Figure 2.4 Percentages of Spectators Who Watched at Least One Sporting Event in Europe.

Source: Based on Eurostat (2016).

An interesting study by Fairley (2009) underlines the role of travel modes for sport event attendees. According to the researcher, a tour bus ride doesn't only provide the access but also plays an important role in the identity of the fan group. During a trip, a good leader can build a travel group that has more effective event participation and contributes to ensuring the sustainability of travel arrangements.

Economic scientists see sport tourism as a new opportunity for global economic development. At the event venue, there is a sales procedure, involving all elements of the sport tourism product. According to researchers, consumer habits influence the whole trip, which increases spending on sports events. The different forms and levels of consumer behavior are influenced by personal, psychological, and environmental factors (Funk, 2008).

The announcement of mega events can be important marketing tools for the host communities, as they act as not only invitations to the competition, but they also invite tourists (Sugden, 2007). For example, each year 130 different locations host mega marathon races, 19 alpine communities organize ski world cups, and many countries compete to win the right to host international championships. Formula 1 host cities even established a network to increase the international tourist arrivals. During these mega events, the atmosphere surrounding the race location is changing, creating an attraction where the whole community will be decorated and traffic rules are changed to secure easier access to the competition site. Host communities also offer accompanying programs, such as happened in Canterbury at the Tour de France event (Bull & Lovell, 2007) or at the 2007 World Cup, where the social factors were the most prevalent, including surprise events (Pettersson & Getz, 2009).

Research Findings on Active Sport Tourism

The definition of active tourism is characterized as physical activity related to traveling sports tourism practitioners, with or without competition (Hinch & Higham, 2001). We are convinced that a sports scientist could contribute greatly, as the diverse field of human movement can be considered as the main field of research in sports science. According to previous study results, all expectations of participating in a sport for a purpose include capacity development. Participants are looking for an optimal location, professional help from which they can sample, and a place where they can experience beneficial environmental impacts and discover the pleasure of movement.

Initially, the study of active tourism products was described as very modest, which was analyzed on the basis of descriptive statistics only (Gibson, 1998). In the early 2000s, however, researchers recognized the importance of these tourism products. Active tourism products were commenced on the basis of influencing factors such as the geographic, socio-economic, environmental, psychological, and behavioral aspects of sports tourism. The results of the research revealed that active sports tourism product is primarily determined by the relationship between the destination and the sports sector (Weed, 2007). Travelers, hosts, and sport providers are important players in tourism product success and have greatly influenced the tourist experience.

Some specialists also feature active sports tourism as part of cultural tourism attractions. At these attractions, the sports cultural background has a great influence on the traveling habits of the participants. In an ethnographic study examining travel motivations, researchers examined tourists migrating abroad to participate in sports tourism events in the Caribbean. Participants commented on the issues that they have been involved in nostalgia during sports events because they want to experience the ambiance of the home environment and are reluctant to spend on the services offered there, helping the local budget (Joseph, 2011). A survey of skiers analyzed the traditions of tourists from different cultures and discovered that cultural identity significantly influenced the motivation and behavior of travelers, highlighting the importance of local services (Chalip & Mc Guirty, 2004; Hudson et al., 2010). Therefore, it is not surprising that foreign tourists are looking for local specialties and a longer stays for their own sports and eating habits. The presence of the usual environment and activities play a major role in motivational factors (Kaplanidou & Gibson, 2010).

Almost all sports tourists are motivated by their own developmental needs, which means they require different conditions based on their different abilities. This has been recognized in many places by sports tourism providers, highlighting the need for various difficulty levels for skiing, mountaineering, and rafting. In a mountain bike contest, the well-prepared and optimal competitive environment is the most attractive factor for travelers who are not competitive athletes (Getz & Mc Connell, 2011). Bikers also consider it very important to have easily accessible racing and training conditions (Buchalis & Darcy, 2010). Testing among the surfers has shown that participants are frequent travelers who attempt the sport on a higher level; they are looking for more challenging races (Sotomayor & Barbieri, 2015). These studies demonstrate that tour operators need to strive for quality conditions and take into account tourist capabilities.

Despite adverse publicity about the negative health effects of sport tourism, sports scientists have repeatedly emphasized the benefits of sport activities, and its promising future (Bodenheimer, 1999; Sallis, 2009). According to a Swiss Health Tourism survey, there is a growing need for healthy lifestyles among tourists, which can be very much influenced by sports activities such as mountain biking, hiking, or playing golf (Laesser, 2011). These tourist activities draw attention to the fact that there are many opportunities for health tourism to connect with sports tourism. In our opinion, there is a great need to involve sport science professionals in order to satisfy these needs. In many cases, sport scientists are able to accurately assess the abilities of active tourists to formulate proposals for loads and to offer capability development programs.

The sports industry can influence tourists, when they provide a well-organized tourism service for their active vacation (Hudson & Hudson, 2015). The effects of activities and services should be analyzed from environmental, social, and economic perspectives in order to understand the sport tourism industry model. There is evidence that factors such as the location, transportation method, and type of sporting event effect traveler attraction. At the same time, there is a lack of sport science approach if sport tourism can contribute to the local athletes' development. For example, it is well known that the popular Austrian ski resort, Annaberg, became one of the most famous tourism destinations because the number one downhill skier, Marcel Hirscher, was born there. Also, the tennis courts in Mallorca are the most booked because Raphael Nadal grew up on the island.

Although personal experiences are often inconceivable and subjective, they are different for every visitor (Highmore, 2002), and, therefore, they do not always provide a realistic view of sport tourism products. This is often related to emotional elements (Kaplanidou, 2010) such as a momentary mood and personal expectations.

Among the dominant branches of active sports tourism, the most popular is bike tourism due to the significant growth of bicycle production and biking trails (Tobin, 1974). Cycling is a potential model of the integration of transport, tourism, and recreation (Lumsdon, 2000). New products emerge within this sector, based on the offers of an extreme environment, and these are based on innovative, challenging activities and impacts. These are ever-growing adventure tourism products whose price categories depend on the characteristics of time, distance, and challenging capabilities (Buckley, 2007).

Researchers clearly state that choosing an active sports tourism destination depends mostly on the climatic conditions. In this context, attention is drawn to the consequences of climate change (Dawson & Scott, 2010; Pütz et al., 2011), such as global warming, which negatively affects the tourism industry at holiday resorts. Due to climate change, the winter sports season could be shorter, which reduces skiing, snowboarding, and outdoor ice sports activities. In contrast, summer resorts increased in summer resorts specializing in winter sports, so their profile had to be changed (Pegg et al., 2012).

According to surveys completed by tourism providers, winter tourism suppliers do not feel threatened by climate change because they are hoping to deal with problems using special technologies (Wolfsegger & Gössling, 2008). Despite this, the Swiss Tourism Organization predict a 4% drop in winter tourism sales by 2030.

Based on another econometric analysis, the knowledge of climate change has had a positive impact on market participants, as both service providers and guests can be prepared for change (Hoffmann et al., 2009).

Publications on the Benefits of Sport Tourism Activities

One milestone in the development of tourism is the appearance of wellness tourism. In our opinion, wellness tourism emerged because the benefits of tourism became the priority in product development and travelers began expecting to feel well from travel. Pampering and therapeutic programs that ensure people's comfort are among many methods that target these goals. As sport professionals, we know that offering physically active programs with high load does not always have pleasant effects, so offering sport touristic programs should be planned and guided very carefully.

The strategic development of the sport tourism industry is influenced by the knowledge of its benefits. A good example of these benefits is in South Africa, where a national sport tourism strategy has been developed to expand potential opportunities because of the positive tourism effects of the FIFA World Cup (Swart & Bob, 2010). With regard to the successes, specialists draw attention to the potential of event legacy. This includes event organization's "know-how", and the infrastructural development. Examination of the heritage of the 2006 World Cup has proved the effects of tourism on the host cities, but some authors point out that comparing events to one another is not always beneficial (Preuss, 2007). Instead of benchmarking, it is suggested to complete a bottom-up analysis to monitor changes in six main elements: infrastructure, knowledge, image, emotions, network, and culture. Many of these elements affect the structure of the host community directly, which can be used to develop long-term benefits.

The influence of sports tourism is widely interpreted by professionals, but many agree that the effects are only worth comparing with other tourism sectors. The social and economic effects are the most common, but changes in cultural identities and social interactions are also experience (Barros, 2007; Hritz & Ross, 2010). Tourists are not just spending money at sporting events, but they also contribute to revitalizing the local economy. Improvement in the local economy is determined by how much visitors enjoy the event, especially when their favorite athletes are involved (Yu, 2010). In the research project conducted on Tour de France visitors, they appreciated the cultural experience most, and also the overall image of participating in such a prestigious event (Balduck et al., 2011). However, visitors complained that in some cases, prices were too high and transportation was too complicated. In Singapore, Formula 1 races are not just organized for sports development but also to be a part of the urban marketing strategy (Henderson et al., 2010).

We consider the model to be familiar with economic impacts, which is not only a matter of direct benefit but also of the consequences of tourist spending (Murray, 2014). Researchers have also repeatedly questioned tourist sporting habits, consumption habits, motivation to purchase sports goods, and how they influence their decision making (Filo et al., 2013). According to the test results, the traveler is influenced by internal and external stimuli for their purchasing decisions, in which individual expectations play a major role. In the event that the traveler spends money on sporting goods, it will in many cases be an investment for a subsequent trip. In the case of service providers and the economic development of the destination, this spending assists in product development, wage growth, and tax revenue increase.

When questioning tourist participation in sporting events, many stated that they will stay on site and spend more money. Conversely, viewers often go home after the event (Scott & Turco, 2007). In examining the costs of professional golfers, the athletes were classified in three categories: low-spenders $69.2, middle-spending $219, and many poets spend $759 daily, on average (Dixon et al., 2011).

As an example of the benefits of sport tourism activities, we can point out that in a popular resort, the local habits and traditions also drive the traveler to get involved in the local lifestyle. At the same time, it is also characteristic that the use of certain services can attract new consumption. For example, if you need a higher intensity load for active sports tourism products, you may, therefore, increase the need for relaxation, liquid, and food consumption. This will also affect the amount of money tourists are willing to pay for the product.

There is also a positive and negative side to the effects of sport activities (Table 2.1), which can lead to a 13–16% increase in tourism spending and in the number of employees in the sector (Hudson & Hudson, 2015). Many studies have been focusing on the multifaceted benefits, but few talk about what the sports industry and society gain as a result of the appearance of sport tourists. We are convinced that hosting sporting events can teach local committees about organizational methods, which can be inspirational for organizing new events.

Table 2.1 The Advantages and Disadvantages of the Sport Tourism Sector in Communities

Advantages	*Disadvantages*
Increasing number of visitors (Preuss, 2005)	Mass tourism effects (Hritz & Ross, 2010)
Larger offer of programs (Schonk & Chelladurai, 2008)	Increasing prices (Smith & Stewart, 2007)
Wider range of consumers (Ritchie & Adair, 2002)	Different behavior (Green & Jones, 2005)
Health benefits (Angus, 2016)	Increased use of health services (Hudson, 2012)
Multilingual development	Miscommunications
Good marketing tool (Henderson et al., 2010)	Increasing air, noise, and light pollution (Wolfsegger & Gössling, 2008)
Appreciation of local values (Fredline, 2006)	Increased sport injuries (De Knop, 2004)
Exploring nature (Standeven & Knop, 1998)	Non-sustainable land use (Dawson & Scott, 2010)
Increasing offers of programs (Gibson, 1998)	
Economic benefits and taxes (Hudson & Hudson, 2015)	
Opportunities for social interaction (Chalip & McGuirty, 2004)	

Active Transportation Compared to Health Indicators (%)

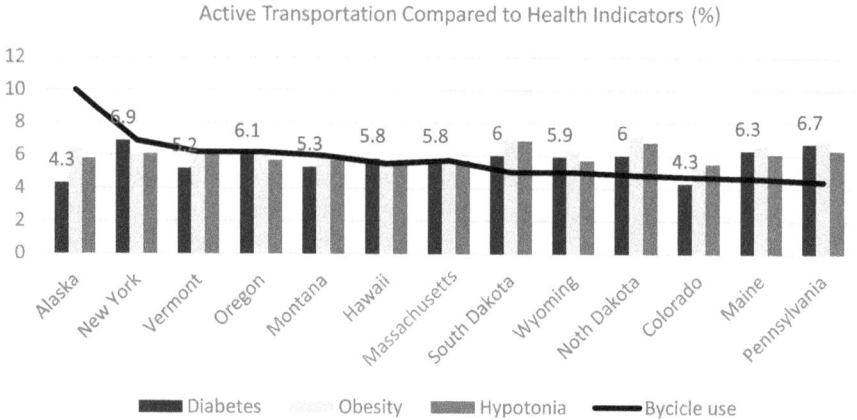

Figure 2.5 Bicycle Use and Health Indicators in Some US States.
Source: Angus (2016).

From a health scientific perspective, the impact of active sport tourism on health is interpreted mostly by comparative health data. In the Netherlands, the analysis of the shift from car to cycling has shown that, although they are more exposed to air pollution and accident hazards, the benefits of physical activity produce 3–14 life-years gained (De Hartog et al., 2010). In another study, the proportion of diseases in the US was compared to the number of pedestrians and cyclists (Figure 2.5) (Angus, 2016). Based on the data, it proves that destinations with more walking or cycling trails available have lower rates of diabetes. Similar results were found in international research examining aggregate cross-sectional health and travel data for 14 countries. The researchers found statistically significant negative relationships between active travel and self-reported obesity. Also, the results have shown significant positive relationships between active travel and physical activity and statistically significant negative relationships between active travel and diabetes (Pucher et al., 2010).

We are convinced that this data can also contribute to the promotion of physical activities for both locals and tourists. These results also demonstrate that tourism should play an increasingly important role in advertising the benefits of physical activity like walking and cycling (Lamont, 2009).

Conclusion

Based on the international literature and research findings from 2005 to 2017, sport tourism can be considered a very complex tourism product. Researchers can analyze sport tourism by highlighting the relationship between the environment and the stakeholders. Writers who have published in the last two decades have built their studies on aspects of the relationship between sport and tourism, which confirms that sport tourism is a multifaceted industry based on sporting activities and attractions. Based on the content analysis of the publications, we can say that researchers have accumulated a number of results that are vital to understanding the industry and the travel habits of those interested.

On this basis, the theory of sports tourism became a well-defined sub-discipline related to tourism. Many scientists were involved in sport tourism studies based on their own questions, assumptions, and methods from their own field of study. Among the scientists, geographers introduced sport tourism from a spatial perspective in their publications, underlining the importance of the environment and the direct and indirect effects of global economic and climate changes.

Several researchers have demonstrated the social effects of sport tourism products from a sociological approach, which, in many cases, are plausible in the development of the host community. Using psychological methods, researchers found that sports travel motivations are mostly aimed at gaining experience, which is influenced by previous experiences and sports profession skills. Results from economic scientists report the economic stimulus effects of sports tourism, which emphasize the global and local importance of sport tourism. In this context, the methodology of sport tourism management has also changed considerably, with the aim of attracting as many stakeholders as possible into the development process.

According to the results of active sports tourism studies, sports tourism is linked to the culture, rules, and traditions of the sport, which mobilizes the characteristics, skills, and knowledge of the traveler. Among the publications, we have found limited information about the intensity of the load among sport tourism activities if they are not a part of a competition. Additionally, there is little information about participants' resting habits in order to explain the reasons for using travel habits and related services more accurately.

We consider inspirational initiatives to address the positive health effects of sport tourism, contributing to an individual's quality of life. We are convinced that researchers further demonstrate the importance of addressing this by contributing to the expansion of physically active programs and services to encourage healthier tourist behaviors.

References

Aho, S. K. (2001). Towards a general theory of touristic experiences: Modelling experience process in tourism. *Tourism Review, 56*(3–4), 33–37.

Angus, H. (2016). *State of the bike walk union: Here are the statistics.* Retrieved March 12, 2018, from https://momentummag.com/state-of-the-bike-union-the-2016-biking-walking-benchmark-report/

Balduck, A. L., Maes, M., & Buelens M. (2011). The social impact of the Tour de France: Comparisons of residents' pre- and post-event perceptions. *European Sport Management Quarterly, 11*(2), 91–113.

Banhidi, M. (2007). Sporttudomány és turizmus [Sport science and tourism.] *Magyar Sporttudományi Szemle, 8*(2), 32–38.

Banhidi, M. (2011). *Sportföldrajz [Sport Geography]* Dialóg Campus Kiadó, Budapest, Pécs

Barros, C. (2007). Analysing the relationship between sports and tourism: A case study of the Island of Madeira. *International Journal of Sport Management and Marketing, 2*(5), 447–458.

Bodenheimer, T. (1999). The movement for improved quality in health care. *The New England Journal of Medicine, 340,* 488–492.

Bonzak, B. (2013). Aktywne formy turystyki – problemy terminologiczne. In R. Wiluś & J. Wojciechowska (Eds.), *Nowe-stare formy turystyki w przestrzeni* (pp. 49–62). ser. "Warsztaty z Geografii Turyzmu", 3.

Buchalis, D., & Darcy, S. (2010). *Accessible tourism, concepts, and issues.* Tonawanda, NY: Channel View Publications.

Buckley, R. (2007). Adventure tourism products: Price, duration, size, skill, remoteness. *Tourism Management, 28*(6), 1428–1433.

Bull, C., & Lovell, J. (2007). The impact of hosting major sporting events on local residents: An analysis of the views and perceptions of Canterbury residents in relation to the Tour de France. *Journal of Sport & Tourism, 12*(3–4), 229–248.

Chalip, L., & Mc Guirty, J. (2004). Bundling sport events with the host destination. *Journal of Sport Tourism, 9*(3), 267–282.

Chen, N., & Funk, D. C. (2010). Exploring destination image, experience and revisit intention: A comparison of sport and non-sport tourist perceptions. *Journal of Sport Tourism, 15*(3), 239–259.

Dawson, J., & Scott, D. (2010). Climate change and tourism in the Great Lakes region: A summary of risks and opportunities. *Tourism in Marine Environments, 6*(2–3), 119–132.

De Hartog, J. J., Boogaard, H., Nijland, H., & Hoek, G. (2010). Do the health benefits of cycling outweigh the risks? *Environmental Health Perspectives, 118*(8), 1109–1116.

De Knop, P. (2004). Total quality, a new issue in sport tourism policy. *Journal of Sport & Tourism, 9*(4), 303–314.

De Pauw, K., Roelands, B., Cheung, S. S., Bas de Geus, B., Rietjens, G., & Meeusen, R. (2013). Guidelines to classify subject groups in sport-science research. *Human Kinetics Journals, 2*, 111–122.

Dixon, A. W., Backman, S., Backman, K., & Norman, W. (2011). Expenditure-based segmentation of sport tourists. *Journal of Sport & Tourism, 17*(1), 5–21.

Edginton, C. R., Banhidi, M., Jalloh, A., Dieser, R. B., Xiafei, N., & Baek, D. Y. (2014). A content analysis of the World Leisure Journal: 1958–2012. *World Leisure Journal, 56*(3), 185–203.

Elsrud, T. (2001). Risk creation in traveling. *Annals of Tourism Research, 28*(3), 597–617.

Eurostat. (2016). *Sport statistics.* Retrieved April 10, 2024 from https://ec.europa.eu/eurostat/documents/4031688/7203321/KS-04-15-823-EN-N.pdf/b911c74d-c336-421e-bdf7-cfcba4037f94

Fairley, S. (2009). The role of the mode of transport in the identity maintenance of sport fan travel groups. *Journal of Sport Tourism, 14*(2–3), 205–222.

Filo, K., Chen, N., King, C., & Funk, D. C. (2013). Sport tourists' involvement with a destination: A stage-based examination. *Journal of Hospitality & Tourism Research, 37*(1), 100–124.

Foo, L. M. (2000). Sports tourism: An Australian perspective. *Tourism Research Report Bureau of Tourism Research, 2*(1), 9–17.

Fredline, E. (2006). Host and guest relations and sport tourism. *Sport and Society, 8*(2), 263–279.

Funk, C. D. (2008). *Consumer behavior in sport and events: Marketing action.* Oxford: Butterworth-Heinemann, Elsevier.

Gaines, C. (2015). *The NFL and Major League Baseball are the most attended sports leagues in the world.* Retrieved from https://www.businessinsider.com/attendance-sports-leagues-world-2015-5

Getz, D., & Mc Connell, A. (2011). Serious sport tourism and event travel careers. *Human Kinetics Journal, 25*(4), 326–338.

Gibson, H. J. (1998). Sport tourism: Critical analysis of research. *Sport Management Review, 1*(1), 45–76.

Gibson, H. J. (2003). Sport tourism: An introduction to the special issue. *Journal of Sport Management, 17*(3), 205–213.

Green, B. C., & Jones, I. (2005). Serious leisure, social identity and sport tourism. *Sport in Society, 8*(2), 164–181.

Henderson, J. C., Foo, K., Lim, H., & Yip, D. (2010). Sports events and tourism: The Singapore formula one grand prix. *International Journal of Event and Festival Management, 1*(1), 60–73.

Highmore, B. (2002). *Everyday life and cultural theory: An introduction.* London: Routledge.

Hinch, T., & Higham, J. (2001). Sport tourism: A framework for research. *International Journal of Tourism Research, 3*, 45–58.

Hinch, T., & Higham, J. (2010). *Sport tourism development.* Tonawanda, NY: Channel View Publications.

Hoffmann, V. H., Sprengel, D. C., Ziegler, A., Kolb, M., & Abegg, B. (2009). Determinants of corporate adaptation to climate change in winter tourism: An econometric analysis. *Global Environmental Change, 19*(2), 256–264.

Hritz, N., & Ross, G. (2010). The perceived impacts of sport tourism: An urban host community perspective. *Human Kinetics Journal, 24*(2), 119–138.

Hudson, S. (2012). *Sport and adventure tourism.* New York, London, Oxford: Haworth Hospitality Press.

Hudson, S., Hinch, T., Walker, G., & Simpson, B. (2010). Constraints to sport tourism: A cross-cultural analysis. *Journal of Sport & Tourism, 15*(1), 71–88.

Hudson, S., & Hudson, L. (2015). *Winter sport tourism. Working in winter wonderlands.* Oxford: Goodfellow Publisher.

Ito, E., & Higham, J. (2020). Supplemental tourism activities: A conceptual framework to maximise sport tourism benefits and opportunities. *Journal of Sport & Tourism, 24*(4), 269–284.

Joseph, J. (2011). A diaspora approach to sport tourism. *Journal of Sport and Social Issues, 35*(2), 146–167.

Kane, M. J., & Tucker, H. (2004). Adventure tourism: The freedom to play with reality. *Tourist Studies, 4*(3), 217–234. https://doi.org/10.1177/1468797604057323

Kaplanidou, K. (2010). Active sport tourists: Sport event image considerations. *Tourism Analysis, 15*(3), 381–386.

Kaplanidou, K., & Gibson, H. (2010). Predicting behavioral intentions of active event sport tourists: The case of a small-scale recurring sports event. *Journal of Sport and Tourism, 15*(2), 163–179.

Kent, M. (1994). *The Oxford dictionary of sports science and medicine.* Oxford: Oxford University Press.

Kyle, D. G. (2015). *Sport and spectacle in the ancient World.* Hoboken, NJ: Wiley Blackwell.

Laesser, C. (2011). Health travel motivation and activities: Insights from a mature market – Switzerland. *Tourism Review, 66*(1–2), 83–89.

Lamont, M. J. (2009). Reinventing the wheel: A national discussion of bicycle tourism. *Journal of Sport and Tourism, 14*(1), 5–23.

Leiper, N. (1979). The framework of tourism: Towards a definition of tourism, tourist, and the tourist industry. *Annals of Tourism Research, 6*(4), 390–407.

Lumsdon, L. (2000). Transport and tourism: Cycle tourism – a model for sustainable development? *Journal of Sustainable Tourism, 8*(5), 361–377.

Murray, A. (2014). *The economic value of tourism.* Retrieved March 17, 2018, from https://www.slideshare.net/mellormurray/the-economic-value-of-tourism

Neuendorf, K. A. (2002). *The content analysis guidebook.* New York: SAGE Publications.

Pegg, S., Patterson, I., & Gariddo, P. V. (2012). The impact of seasonality on tourism and hospitality operations in the alpine region of New South Wales, Australia. *International Journal of Hospitality Management, 31*(3), 659–666.

Pettersson, R., & Getz, D. (2009). Event experiences in time and space: A study of visitors to the 2007 World Alpine Ski Championships in Åre, Sweden. *Scandinavian Journal of Hospitality and Tourism, 9*(2–3), 308–326.

Pigeassou, C. (2004). Contribution to the definition of sport tourism. *Journal of Sport & Tourism, 9*(3), 287–289.

Preuss, H. (2005). The economic impact of visitors at major multi-sport events. *European Sport Management Quarterly, 5*(3), 281–301.

Preuss, H. (2007). FIFA World Cup 2006 and its legacy on tourism. In *Trends and issues in global tourism 2007* (pp. 83–102). Berlin, Heidelberg: Springer Berlin Heidelberg.

Pucher, P., Buehler, R., Bassett, D. R., & Dannenberg, A. L. (2010). Walking and cycling to health: A comparative analysis of city, state, and international data. *American Journal of Public Health, 100*(10), 1986–1992.

Pütz, M., Gallati, D., Kytzia, S., Elsasser, H., Lardelli, C., Teich, M., Waltert, F., & Rixen, C. (2011). Winter tourism, climate change, and snowmaking in the Swiss Alps: Tourists' attitudes and regional economic impacts. *Mountain Research and Development, 31*(4), 357–362.

Ritchie, B., & Adair, D. (2002). The growing recognition of sport tourism. *Current Issues in Tourism, 5*(1), 1–6.

Ritchie, B. W., & Adair, D. (2004). *Sport tourism interrelationship, impacts and issue.* Tonawanda, NY: Channel View Publications.

Ross, S. D. (2001). *Developing sports tourism. An eGuide for destination marketers and sports events planners.* Retrieved February 15, 2018, from http://www.lib.teiher.gr/webnotes/sdo/Sport%20Tourism/Sport-Tourism%20Development%20Guide.pdf

Sallis, R. E. (2009). Exercise is medicine and physicians need to prescribe it! *British Journal of Sport Medicine, 43*(1), 3–4.

Schonk, D. J., & Chelladurai, P. (2008). Service quality, satisfaction, and intent to return in event sport tourism. *Human Kinetics Journal, 22*(5), 587–602.

Scott, A. K. S., & Turco, M. D. (2007). VFRs as a segment of the sport event tourist market. *Journal of Sport & Tourism, 12*(1), 41–52.

Slack, T., & Hinings, B. (1987). Planning and organizational change: A conceptual framework for the analysis of amateur sport organizations. *Canadian Journal of Sport Science, 12*(4), 185–193.

Smith, A. C. T., & Stewart, B. (2007). The travelling fan: Understanding the mechanisms of sport fan consumption in a sport tourism setting. *Journal of Sport & Tourism, 12*(3–4), 155–181.

Sotomayor, S., & Barbieri, C. (2015). An exploratory examination of serious surfers: Implications for the surf tourism industry. *International Journal of Tourism Research.* https://doi.org/10.1002/jtr.2033.

Standeven, J., & Knop, P. (1998). *Sport tourism.* Champaign: Human Kinetics Publisher.

Starr, M. (2000). *The effects of weather on retail sales.* Available at SSRN 221728.

Sugden, J. (2007). Running Havana: Observations on the political economy of sport tourism in Cuba. *Leisure Studies, 26*(2), 235–251.

Swart, K., & Bob, U. (2010). The eluding link: Toward developing a national sport tourism strategy in South Africa beyond. *South African Journal of Political Studies, 34*(3), 373–391.

Sydney Sports and Exercise Physiology. (2010). *What is sport science?* Exercise and Sport Science Australia. Retrieved from https://www.ssep.com.au/what-is-sport-science.html.

Tobin, G. (1974). The bicycle boom of the 1890s: The development of private transportation and the birth of the modern tourist. *Journal of Popular Culture, 8,* 838–849.

Turco, D., Riley, R., & Swart, K. (2002). *Sport tourism.* Morgantown: Fitness Information Technology, Inc.

Wang, D. (1998). On the disciplinary nature of tourism science. *Tourism Tribune.* Retrieved from http://en.cnki.com.cn/Article_en/CJFDTOTAL-LYXK802.013.htm

Weed, M. (2007). *Sport & tourism: A reader.* London: Routledge, Taylor and Francis Group.

Wittmann, L. (2017). *Sport tourism: Sleeping giant of the tourism market.* Retrieved March 20, 2018 from https://www.skal.org/sites/default/files/media/Public/Web/PDFs/sporttourism.pdf

Wolfsegger, C., & Gössling, S. (2008). Climate change risk appraisal in the Austrian Ski industry. *Tourism Review International, 12*(1), 13–23.

Yu, C. C. (2010). Factors that influence international fans' intention to travel to the United States for sport tourism. *Journal of Sport & Tourism, 15*(2), 111–137.

Part II
Sport Tourism Destinations

Photo: Giyasettin Demirhan

3 Sport Traditions, Habits, Rules, and Policies

Examples from the World

Bijen Filiz and Giyasettin Demirhan

Sport Traditions, Habits, Rules, and Policies

At the beginning of human existence, sport was not for the purpose of protecting people's health and beauty, using their excess energy, evaluating their leisure time, contributing to peace, or providing commercial benefits. Sports began for the purpose of developing the body and muscles to protect from the natural conditions of human nature. In the following periods, sport, which develops through social interaction in society, influenced the ideas and behaviors of the society and, therefore, the elements of the culture (Güven, 1999).

Sport affects almost every aspect of social life and has developed into different branches based on the characteristics of materials used in everyday life and the coins since it started to be seen in the history scene. Therefore, the sports performed in a particular society consist of the conditions that it possesses such as technology and infrastructure (Demirbolat, 1988). After people started to live together by forming communities, they gradually formed social relationships and the process of social influencing created values, rules, management, and lifestyles. Therefore, the influence of traditions on sport has continued with different styles and effects throughout the ages (Güven, 1999).

Sports like American football and baseball in the United States of America (US), football and rugby in England, judo in Japan, and taekwondo in South Korea have been adopted as a product of the culture in these societies. Golf in the US and England, winter mountain sports in Central and Northern Europe, and cycling in France are performed traditionally and commonly, and countries like Canada and Japan are following suit. Considering that, in Turkey, sport tourism is a newly emerging concept, and that Turkey has a great potential in sport tourism, it is very important to involve sport tourism fields in tourism. Turkey, surrounded by sea on three sides, is a very rich country in terms of its natural, historical, and cultural values. Due to its geographical location, it has four seasons and, therefore, each region from north to south and from east to west has opportunities for different sport fields. It is possible to perform various sport activities in different regions of Turkey, such as mountain climbing, mountain biking, hiking in the Black Sea and Central Anatolia regions, winter sports in Northeastern Anatolia and Marmara region, free-style jumping, paragliding, rafting, trekking, skiing, water ski, and

DOI: 10.4324/9781003476658-5

hunting in Mediterranean and East Anatolian regions (Anvar, 2022). There is also the opportunity for traditional sports like wrestling and oil wrestling which are performed in the Mediterranean, Thrace, and Marmara regions. In addition, Turkey has sufficient opportunities for organizations in sport branches like football, basketball, volleyball, tennis, and golf due to its geographical location and human resources. Importance is given to the sustainability of Turkey's sport infrastructure improvement policy, with completed and ongoing construction of sport facilities in many different regions (Sümer, 1989).

Through these examples, sport habits have strengthened the significance and place of sport tourism and provided substantial pecuniary resources to the countries. Accepting the fact that sport has a positive influence on society has led to regarding sport tourism as a public service for public interest in developed countries. The cooperation and action processes realized in sports fields with ranked management systems also facilitated expanding the sport to target groups. Thus, sport activities, with the tourism sector increasing day by day, have become traditional as a result of performed long-lasting successful sport policies.

The influence of cultural structure on sport is mainly due to values, traditions, customs, and cultural changes. Values, judgments, traditions, and customs that emerge in parallel with a society's values system can affect an individual's sport participation in a positive or negative way by influencing the cultural structure of what people do and how they do it (Yetim, 2006). All efforts can only be successful in spreading and embracing the sport if it is suitable for the cultural system in that particular society. Sport, which is also a demonstration of the level of societal cultural development, guides the lives of people directly or indirectly in social life and takes more and more space in their everyday lives.

Sports started to strengthen its place in the life of mankind and was promoted in the West through the transfer of knowledge and culture after the industrial revolution. After the revolution, standard rules had emerged and professional organizations and international competitions have been organized. In our current period, sports are increasingly divided into branches of expertise (Yıldız & Çekiç, 2015). After the first steps of globalization, international federations were established. In 1863, the Football Association was established in England to develop new types of football. During these years, rugby and American football developed in America, causing modern sports to increase all over the world. England had also modernized sports from other countries, such as tennis, of which is based on the French Renaissance (Ogrozio, 2006).

Sport Tourism and Examples from the World

The concept of sport has begun to gain importance in every area of the world and is now progressing toward becoming an important industry that is complementary to tourism. The roots of sports tourism, based on sports cultures and habits, span back to the trips made in ancient Greece to participate in the ancient Olympics. In this regard, sports tourism can be explained as traveling to the national or international arena in order to participate, observe, or just to be present (Ataçocuğu, 2008).

The primary goal of sports tourism is to contribute to the socio-economic and cultural values of a place by traveling to other regions, to participate in sport activities, to watch sports (İçöz & Kozak, 2002).

The fastest-growing area within the tourism industry is travel related to sports and physical activity. According to a survey, people expressed that the opportunities offered for participation in sporting events are decisive in holiday preferences. To support this idea, the global relationship between tourism and sport has started to increase with momentum in recent years; this relationship has qualified to continue into the new century (İçöz & Kozak, 2002).

Sports around the World

In many countries, activities within the scope of sports tourism attract many tourists. In fact, sports tourism activities in some countries have become popular around the world and have become identified with the region. Especially in countries, such as America, Japan, Canada, and New Zealand, sports tourism has developed quite a lot and many activities related to sport tourism exist in these regions. Sports events identified with traditions, such as horse tourism, canoeing, canyon tourism, and American football in America; mountain biking, trekking, and mountaineering in New Zealand; fishing, horse riding, and judo in Japan; and skiing, canyon tourism, and canoeing in Canada all have a large share in sports tourism (Gibson, 1998).

Depending on the geographic location, and cultural characteristics that constitute social construction, societies give preference to different sports branches. For example, hockey is an interest in Asian countries, taekwondo in South Korea, and wrestling in Turkey is more prevalent compared to other countries. Additionally, golf is one of the most popular sports, especially in developed countries. Golf is the fourth most popular sport in the world with over 60 million golf players and approximately 36,000 known golf courses. The US is ranked first as the country with the most golf players and countries in Asia are ranked second in the world. There are 1,849 golf courses and over 678,000 golfers in the UK, one of the world's leading destinations. Germans and Swedes follow the British in the rank of passion for golfing, and while there are 637,000 golfers in Germany and 473,000 in Sweden, these countries are followed by France, Scotland, Ireland, Spain, and Portugal (Strick, 1987). On the other hand, ice hockey is among the most popular sports in America and Canada. In addition, Russia, Scandinavian countries, and Central European countries follow this order.

In regions such as South America, the Philippines, Japan, and Russia, boxing is loved both at amateur and professional levels; this is why many good boxing athletes emerge from these countries (Ogrozio, 2006). The game of rugby has been enthusiastically followed by countries of the former British Empire. Therefore, rugby has become very popular in these countries due to its difficulty and the spirit of the sport. Even its derivatives such as American Football and Australian Football have created a different field for themselves. Rugby, which is played by most of the UK, Australia, South Africa, and some European countries, hosts very important organizations, especially national teams. These rugby organizations are also increasing the popularity of sports in these countries (Strick, 1987).

Cricket, another sports branch, is being followed by large masses, especially in the countries of the Old English Empire. This sport is admired because it carries cultural features of the UK. Cricket is played at the highest level both as national teams and club teams in countries such as India, Pakistan, Bangladesh, England, and Australia. In some countries, some sports branches gain national qualities, that is to say, sports are unique to that country and form the culture of that country.

Cycling, as a sport, has a very important position in developed countries (Passafaro et al., 2014). It is recorded that, in 1996, 56,000 people participated in bicycle tourism in Denmark. Other developed countries have important cycling events such as Le Tour, a high-altitude three-week bike race in France, one of the world's major sporting events.

Sports in Turkey

Considering that sports tourism is an emerging concept of tourism in Turkey, it has a great potential. Turkey, surrounded by sea on three sides, is a very natural and rich country in terms of historical and cultural values. Due to its geographical location, it has four seasons, this means there are separate sport tourism opportunities from north to south and east to west (Bektaş, 2010). Sports tourism has begun to take its place among other types of tourism in Turkey and it is possible to say that it would be permanent, due to the advantages Turkey has (Bektaş, 2010).

Sporting activities, which have become increasingly related to tourism in recent years, have emerged as an area with growth potential. The growth potential of sport tourism made it necessary to develop substructure and superstructure to host the sports and gain a significant share of this rising market. In Turkey, sports tourism is seen as an important type of tourism. This explains why there are 178 tourism area for outdoor sports, among which are 23 mountaineering, 16 snow sports, 8 rafting, 4 paragliding, 31 nature walking, 10 canyoning, and 11 diving areas (Bektaş, 2010).

Turkey, in particular, has sufficient opportunities in sport branches relative to its physical geography and human participation in football, basketball, volleyball, tennis, and golf and the most followed sports in the world football, volleyball, basketball in terms of audience. These sports organizations have begun to improve themselves with ongoing and completed sports facilities in many different regions. The type of facilities that serve the general tourist population in Turkey are mainly concentrated in Istanbul and Antalya in parallel to their performance throughout the tourism industry (Figure 3.1).

Turkey ranked 14th on the Global Sport Index based on the amount of present and future sporting events hosted, holding 14 events of global magnitude in the last eight years. Total of sports tourism in Turkey is found at the level of 1.5%, golf and football constitute the largest part of the share (Bektaş, 2010).

The Effects of Sports Organizations in the World on Tourism

It is clear that sports preferences strengthen the importance and physical location of sports tourism in the world and provide substantial financial resources to the host countries. Economic mobility is experienced in the city through sports tourism

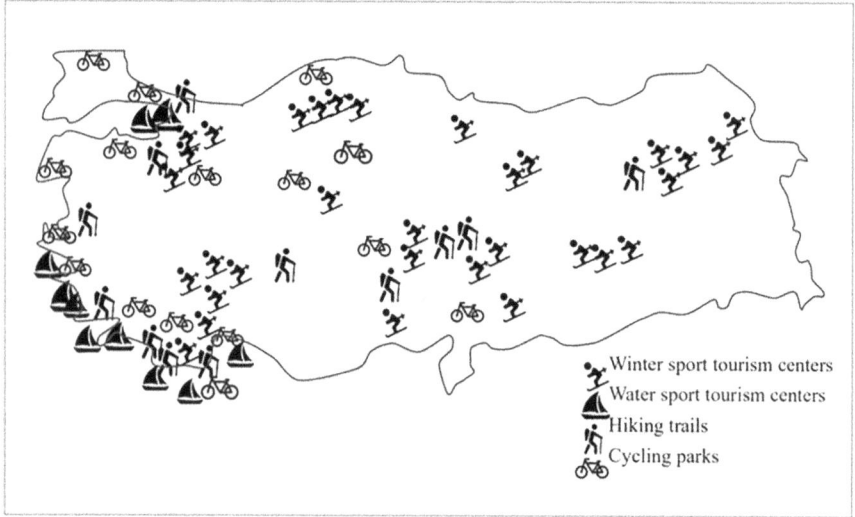

Figure 3.1 Main Sport Tourism Destinations in Turkey.

thanks to tourist attendance at sports organizations, bringing vitality to the country in an economic sense. Large-scale events, such as the Olympic Games and the FIFA World Cup, attract considerable attention. For example, the Super Bowl tournament in Atlanta in 2000 attracted more than 1 million tourists. In light of these facts, investment in sports tourism is also increasing. An example, Walt Disney invested $200 million USD in the "Wide World of Sports" complex in Orlando, Florida. In addition to this sport complex in Disney, the park-themed "Olympic Spirit" was established in Munich, Germany (Gammon & Robinson, 2003).

The field of sports also activates the tourism industry. Atlanta, Georgia, which had previously been an ordinary, poor, and even crime-ridden American city, became one of the world's leading cities after it organized the Summer Olympic Games in 1996. After being named the host of the Olympics, the number of tourists who visited the city increased and so did the number of touristic organizations in the city (Tumer & Rosentraub, 2002). Again, at the Barcelona Summer Olympics in 1992, the city escaped from its previous distorted image by organizing the Olympic Games and becoming a tourist city. Moreover, cities that hold these Olympic sports facilities have earned the reputation of becoming a destination that attracts sports tourists.

Apart from the economic effects of the tourism-sport relationship, there are many additional positive contributions. These contributions include:

- an increase in the number of sportive activities in the country each year,
- the development of professionalism in the sport,
- the increased participation in the sport,
- contributing positively to sportive efficiency in certain regions of the country,
- attracting more national press.
- sharing their own culture with the whole world (Kozak, 1998).

The intense interest and reputation at the social level of sport, the lasting values learned by participants, and the ability to overcome language barriers have made sport tourism a competitive advertising tool for every society. Large events such as the Olympics and global championships also provide an opportunity for participants to show that they are better and more successful than other cities and countries that have previously hosted organizations and competitions. It is possible to say that sport has now become a tool for propaganda and advertising, far beyond being an activity for leisure or a healthy lifestyle. In order to make the best use of this vehicle, federal and local governments grant space in their work programs and devote a significant portion of their investments to sports (Yetim, 2006).

It seems that countries and their respective cities are almost in a race to host sports organizations by the investments they make, as well as the lobbying and promotional activities they conduct. Organizing successful large-scale sporting events and gaining success in international sport competitions are important indicators of national dignity (Ataçocuğu, 2008).

Countries are struggling with all infrastructures, organizations, and economic forces in the international arena to host giant organizations such as FIFA World Cup, Olympic and Paralympic Games, Mediterranean Games, European Football Championship, World Wrestling Championships, World Basketball Championships, and so on. Because hosting the Olympics and world championships is increasing in popularity, investing in cities and strengthening sports infrastructure contributes to the promotion of the country and offers an important urban development opportunity. Such organizations provide significant revenue sources throughout the year as well as reviving the tourism sector by improving the accommodation and food beverage industry. Hosting these events also provides the benefits of international relations, such as sponsorship, infrastructure, facilities, and commercial product sales, as such organizations are national indications of reputation and prestige. Such large organizations offer the host city very important publicity opportunities, and the presence of these international standard facilities has a positive impact on their future organizational candidacy. Today, Barcelona is still preferred by international sporting organizations and sports tourists thanks to the sports facilities inherited from the Olympic Games.

The sports sector is increasingly attracting worldwide public attention as sponsorship and media investments in this area are growing. In the European Union countries, sport is carried out with a well-structured and organized system which contributes to recognizing its own values and customs. Despite this, it is fairly new in the European Union to address politics in the field of external sport relations. Therefore, large-scale sporting events and competitions have strong potential for the development of sports tourism in Europe (Börzel, 1997).

The Europe 2020 Strategy is the European Union's growth strategy. This strategy aims to create smart, sustainable, and inclusive growth in the European Union under changing global conditions (European Commission, 2010). Although sports are economically important, a large part of sports activities take place within non-profit structures. It is not easy for these types of activities to be financially sustainable; therefore, it is important to strengthen the financial support among sports professionals.

However, the role of the European Union in foreign relations becoming increasingly strong over time due to the fact that sports are supportive tools of education, health, intercultural dialogue, and peace development. It is clear that sport policies developed in the European Union will directly contribute to worldwide sport tourism. These sport policies and health awareness will act as a source of great motivation for people and will increase individual and group participation in sports tourism. The creation of new markets and the expansion of existing markets will bring new demands and new infrastructures. The increase in the popularity and magnitude of long-distance travels will also benefit the growth of less developed industrial societies (Fişek, 1989). Therefore, thanks to properly implemented and sustainable sports policies, the number of sports tourists will increase alongside their participation rates.

Conclusion

It can be said that sports tourism is increasingly important in the tourism sector, rapidly spreading to every corner of the world, and reinforcing its position in countries by growing with the sports traditions, habits, and rules. As mentioned in the related parts of the study, whether as a primary reason for sports tourism, or as a derivative of sporting events, sports tourism is an important tourism movement that has attracted millions of people around the world. For this reason, sports tourism has a large environmental influence by creating very important social and economic effects. In addition, sports tourism has become an extremely important community by promoting tourism movements created by sports events and by improving a region's prestige. Therefore, it is popular for tourism activities to be integrated with the well-known cultural characteristics of a country or city.

The sports movement shows us how important its impact on sports tourism is, such as expanding the business volume in the economy, revitalizing the sectors, intensifying commercial transactions, and contributing to the development of physical and institutional infrastructure in the country. Sporting events such as tournaments, world championships, and Olympic Games play an important role in promoting the geography, climate, architecture, art, music, food habits, attractions, and history of the host location and, therefore, constitute the building blocks of sports tourism. Athletes and audiences from many different societies will have the chance seeing new countries, share cultures through intercultural dialogue, representing their own cultures, and have the opportunity to develop in a multicultural manner. This exposure will allow people to share culture with future generations through experiences of sport traditions, habits, rules, and policies.

References

Anvar, A. (2022). *12 adventure sports in Turkey that you must try on your next visit in 2022.* Retrieved from https://traveltriangle.com/blog/adventure-sports-in-turkey.

Ataçocuğu, M. Ş. (2008). *Olimpiyat oyunlarının ev sahibi ülke ve kente ekonomik etkilerinin araştırılması [Investigation of the economic effects of the Olympic games on the host country and city].* [Unpublished master's thesis] Muğla University, Muğla.

Bektaş, F. (2010). *Kaçkar havzası trekking parkurlarının spor turizmi bakımından değerlendirilmesi [Evaluation of trekking trails in terms of sports tourism in Kackar catchment].* [Unpublished doctoral dissertation] Gazi University Education Sciences Institute, Ankara.

Börzel, T. A. (1997). What's so special about policy networks? – An exploration of the concept and its usefulness in studying European governance. *European Integration Online Papers*, 1.

Demirbolat, A. (1988). *Toplum ve spor*. Kadıoğlu Matbaası.

European Commission. (2010). *Communication from the commission Europe 2020*. Retrieved from https://ec.europa.eu/eu2020/pdf/COMPLET%20EN%20BARROSO%20%20%20007%20-%20Europe%202020%20-%20EN%20version.pdf

Fişek, K. (1989). *Spor yönetimi [Sport management]*. A.Ü.S.B.F.

Gammon, S., & Robinson, T. (2003). Sport and tourism: A conceptual framework. *Journal of Sport Tourism, 4*(3), 21–26.

Gibson, H. (1998). Active sport tourism: Who participates? *Leisure Studies, 17*, 155–170.

Güven, Ö. (1999). *Türklerde spor kültürü [Sports culture in Turks]*. Atatürk Kültür Merkezi Başkanlığı.

İçöz, O., & Kozak, M., (2002). *Turizm ekonomisi [Tourism economy]*. Turhan Kitapevi.

Kozak, N. (1998). *Otel işletmeciliği: Kavramlar-uygulamalar [Hotel management: Concepts-applications]*. Turhan Kitapevi.

Ogrozio, R. (2006). *Les enjeux liés aux musées du sport l'exemple Italien*. Musée du Sport.

Passafaro, P., Rimano, A., Piccini, P. M., Metastasio, R., Gambardella, V., Gullace, G., & Lettieri, C. (2014). The bicycle and the city: Desires and emotions versus attitudes, habits and norms. *Journal of Environmental Psychology, 38*, 76–83.

Strick, D. (1987). *Golf: The history of an obsession*. Phaidon.

Sümer, R. (1989). *Sporda demokrasi [Democracy in sport]*. Şafak Matbaa.

Tumer, R. S., & Rosentraub, M. S. (2002). Tourism, sports and the centrality of cities. *Journal of Urban Affairs, 25*(5), 487–492.

Yetim, A. (2006). *Sosyoloji ve spor [Sociology and sport]*. Morpa Kültür Yayınları.

Yıldız, Z., & Çekiç, S. (2015). Sport tourism and its history and contribution of Olympic Games to touristic promotion. *International Journal of Science Culture and Sport, 4*, 326–337.

4 Environmental Aspects of Sport Tourism and Recreation

Monir Shahzeidi and Farhad Moghimehfar

Introduction

Over the past 60 years, different aspects of the environmental impacts of tourism and recreation have been studied. Recreation ecology has emerged as a response to the fast-growing tourism and recreation sectors and as a response to the question of 'whether tourism and recreation can be managed as sustainable sectors of the economy' (Monz et al., 2013). Researchers such as Wagar (1964) established the foundation of tourism and recreation ecology, and their work was followed by other researchers (e.g., Cole, 1989; Hammitt et al., 2015; Hammitt & Cole, 1998; Liddle, 1991) to the extent that, by the early 2000s, thousands of studies had been published in this field (Buckley, 2005; Hammitt et al., 2015; Marion et al., 2016). Although it began with the attention of recreation ecologists, environmental impacts of sport tourism and recreation have sparked social researchers' interest. Social scientists were attracted to this field in the 1970s when air pollution, water pollution, and energy consumption were emphasized as key elements of concern by the United States' National Policy Agenda (Stefanic & Delgado, 1996). In this agenda, ecologists and social researchers were invited to develop tools that measure the environmental impacts of activities including tourism and recreation (Dunlap, 2008). Several tools and measures were created which resulted in the publication of numerous studies. Through the validation of these tools and survey instruments, our understanding of the environmental impacts of tourism and recreation increased. Impacts of sport tourism, including recreational and professional sport as well as outdoor recreation, have been studied carefully.

Researchers described a mutual relationship between environment, recreation, and sport tourism activities. The quality of the environment is dependent upon the type of use and management of sport tourism and recreation events; the quality of the sport tourism and recreation experience is also dependent upon the quality of the environment in which the event takes place (Mason, 2015). This relationship, however, has changed overtime. Traditional approaches exploited the natural environment for the sake of recreation and sport. More recent planning approaches consider a mutual relationship between the environment and sport tourism. This chapter explains this mutual relationship, where both sides benefit from each other, and tourism uses and protects the environment as it is the context where the activity takes

DOI: 10.4324/9781003476658-6

place. First, we will define the environment. Then, we will discuss the positive and negative environmental impacts of sport tourism and recreation. Finally, we will examine two of the most effective environmental impact management frameworks.

What Is the Environment?

The environment provides the physical context for sport tourism and recreation. The concept of environment is not limited to the natural environment. Researchers defined four different environments that host various forms of tourism and recreation (Mason, 2015; Swarbrooke, 1999; Figure 4.1):

- *The natural environment* which includes all the naturally occurring living and non-living elements. The natural environment includes but is not limited to mountains, seas, rivers and bodies of fresh water, caves, deserts, and woodlands.

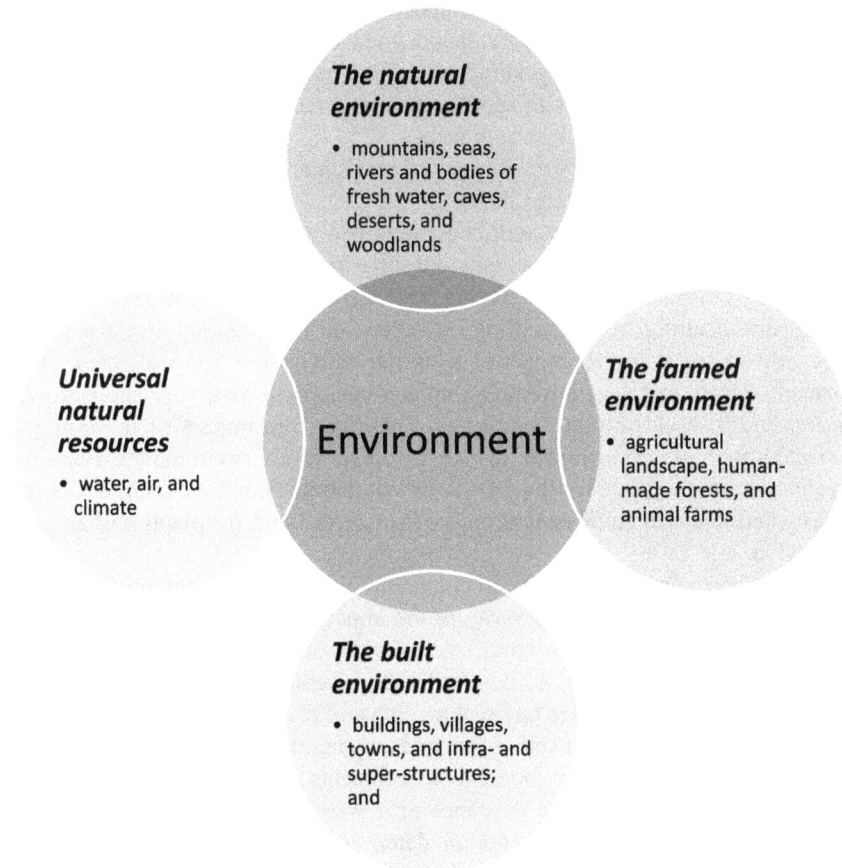

Figure 4.1 The Four Environments.

Source: Adapted; Mason (2015).

- *The farmed environment* includes all the agricultural landscape, human-made forests, and animal farms.
- *The built environment* includes all human-made structures such as buildings, villages, towns, and infra- and super-structures.
- *Universal natural resources* including water, air, and climate.

Although sport tourism and recreation impact all four environments, the focus of this chapter will be on the natural environment and natural resources. In the next sections, we will present different approaches to studying the environmental impacts of sport tourism and recreation. Then, the positive and negative environmental impacts will be defined and explained.

Classification of Sport Tourism and Recreation Impacts

Several different environmental impacts have been associated with sport tourism and recreation activities. To be able to understand, analyze, and manage these impacts, we need to classify the factors that define the impacts, and identify links and connections with the activities and causes of the impacts. Researchers have classified environmental impacts of sport tourism and recreation in many ways. This chapter highlights three of these approaches.

The first approach studies the impacts of tourism and recreation based on three different perspectives (Buckley, 2004):

Type of ecosystem. The type of ecosystem is a significant determinant of environmental impacts of sport tourism and recreation activities (Lakhouit, 2019). Similar activities have different impacts in different environments. For instance, the impact resulting from trampling on vegetation in an upland meadow is significantly higher than the impact of a similar activity in a tropical forest environment. Plants and vegetation in an upland meadow have a very short growth period due to high altitude conditions. Therefore, the impact of trampling is considerably more compared to that in a tropical forest environment where the weather is more suitable for the growth of vegetation. Another example is desert areas where the arid environment makes it more difficult for plants and animals to survive.

Type of activity. The type of sport tourism and recreation activity is also an important determinant of the severity of the impacts (Buckley, 2011). For example, the impact resulting from motorized activities such as off-road motorcycling is more significant than that of hiking. Similar patterns exist for activities such as motorized boating as opposed to kayaking. Several research studies compared the impact of these activities on soil, water, and vegetation and verified differences in the severity of the impacts on different environments.

Management regimes. The existence of a systematic monitoring and impact management system is also an important determinant of the severity of impacts of sport tourism and recreation activities. In remote areas, for instance, where consistent impact monitoring is difficult, management of impacts is more difficult.

Mason (2015) classified factors that define impacts of sport tourism and recreation into six different categories:

Type of activity. Similar to Buckley's (2004) classification, the type of activity is considered as an important determinant of sport tourism and recreation impacts in this approach. For example, the impacts of mountain biking on soil and vegetation are significantly higher than that of road biking. Similar patterns exist when we compare the impacts of motorized boating with kayaking.

The place that the activity occurs. The impacts of sport tourism and recreation is different in different places. For example, the impact of hiking on sand and soil is more significant compared to hiking on a rocky surface or snow. The second principle of the Leave No Trace framework follows this advice: travel and camp on durable surfaces (Leave No Trace Center for Outdoor Ethics, 2020).

Scale and duration of the activity. This factor refers to the number of participants in the activity. For example, mega events and popular activities have greater environmental impacts compared to smaller scale events. Also, whether the environment in which the activity takes place is a permanent host, or a temporary destination, is another important determinant of impact.

The nature and availability of infra-structure. The environmental impacts of sport tourism and recreation activities is significantly lower in places with more infra- and super-structure available. Also, the quality and capacity of the infra- and super-structure determine the number of people the place can host.

Time and seasonality. The time of the activity is very important. For instance, sport tourism and recreation activities, such as hiking, have less impact during the snow season. Similar patterns exist for activities such as downhill skiing when compared to downhill biking in snow resorts during the summer season.

Ecosystem components. Ecosystem components such as flora and fauna are sensitive to changes resulting from sport tourism and recreation activities. Ecosystems that host endangered species are more vulnerable to impacts of sport tourism and recreation.

Direct and Indirect Impacts

Another way to view the impacts of sport tourism and recreation is the direct and indirect influence of such activities on the environment (Magro-Lindenkamp & Leung, 2019). Direct impact refers to obvious and tangible changes that occur as the result of sport tourism and recreation activities. Soil compaction, erosion, deforestation, littering, and air pollution are examples of direct impacts. Indirect impacts are less obvious and more difficult to manage. Sound pollution and its impact on the mating behaviors of animals is an example of indirect impacts. Another example is pathogens that are introduced into an area as the result of sport tourism and recreation activities. Often, indirect impacts are neglected. Therefore, the impact becomes severe and difficult to manage.

Negative and Positive Impacts of Sport Tourism and Recreation

Throughout history, humans have impacted the natural environment through migration, trampling, building structures, hunting, domestication of animals, as well as sport tourism and recreation activities. The impact is, however, more significant in the past century due to population growth. Also, increased leisure time as a result of more structured work and lifestyle change has resulted in more time spent outdoors participating in sport and recreation activities (Li, 2008). Therefore, the impact of human activities has been significant. On the other hand, human's knowledge of the environment has dramatically increased over the past century. Therefore, we have been able to monitor impacts and create positive impacts in some cases. This next section highlights different negative and positive impacts of sport tourism and recreation. Then, a few of the most effective tools created to control the negative impacts of sport tourism and recreation activities will be introduced.

Negative Impacts

Impacts on Vegetation

As one of the most systematically studied topics in sport tourism and recreation, impacts of trampling on vegetation has been investigated in numerous surveys and experiments. These studies were conducted in controlled and uncontrolled environments. Researchers identified type of use, area of use, amount of and intensity of use, user behavior, time of use, and durability of the trampled surface as the most important determinants of the impacts on vegetation (Cole, 2004; Pickering & Hill, 2007; Sun & Walsh, 1998). Most of these studies compared the surviving cover on the trampled area with the initial cover on the area.

Impacts on Soil

Similar to other impacts, the impact on soil is dependent on the area, activity, and management (Cole, 2004; Godtman Kling et al., 2017). Generally, managing the impacts of sport tourism and recreation on soil is more difficult in remote areas, where it is more challenging to manage the impacts and people's behaviors. Mason (2015) believed that in remote areas, management is more complex, knowledge about the area and visitors' behavior is limited, engineering of the area is more difficult, and separating the natural environment, where sport tourism and recreation occur, from the surrounding environment is burdensome. The severity of the impact of sport tourism and recreation on soil is also dependent on climate-related factors, such as precipitation and temperature. These factors often increase or decrease the impact of recreation on soil. For example, in areas with a long snow season, people participate in activities, such as skiing and snowshoeing, that are less impactful compared to mountain biking and hiking. Another example is the higher rates of erosion in areas with greater yearly precipitation. Aspects of soil structure, such as texture and chemistry, are also very important factors.

Impacts on Water

One of the most significant environmental impacts of sport tourism and recreation is on water (Fox, 2019). This is due to many sport tourism and recreation activities being water-based. Whether motorized or non-motorized, these activities have considerable impact on water. Impacts include hydrocarbon releasing into the water, oil, noise, grey water, and the antifouling paints used to protect vessels. Motorized activities, however, have shown to be more impactful. Mosisch and Arthington (2004) identify three categories of impacts resulting from motorized water-based sport tourism and recreation activities: impacts as a result of propeller action (e.g., turbulence and disturbance of sediments), wash (e.g., washing out of roots, damage to the riparian vegetation), and direct contact (e.g., bank erosion). With a different perspective, Mosisch and Arthington used direct and indirect impacts to categorize these damages. Noise disturbance, pollution, nest disturbance, injury, or death from direct contact are examples of direct impacts; change in habitat, change in the PH of the water, migration of water creatures, and reduction in food resources are indirect impacts. Other impacts such as the introduction of plants and pathogens are also indicated in the literature. In addition to these impacts, sport tourism and recreation facilities, such as golf courses, have a significant impact on water. Most of these golf facilities and resorts require a considerable amount of water. Also, the fertilizers used to improve the esthetics of the landscape degrade the water quality in surrounding areas (Petrosillo et al., 2019).

Impacts on Wildlife

As a component of ecosystems, wildlife is vulnerable to excessive sport tourism and recreation activities. Green and Giese (2004) highlighted a range of negative impacts on wildlife. They placed impacts into three different classifications: short-term changes, long-term changes, and impacts at the ecosystem level. Short-term impacts refer to impacts that occur shortly after the presence of sport tourism and recreation participants. Examples of these impacts are changes in the physiology of the wildlife due to changes in food, behavior, and habituation of animals. Long-term impacts reflect effects that are not obvious at the first stages of sport tourism and recreation development. The significance of these impacts increases over time. Examples of long-term impacts are increases in the mortality of certain species and reduced breeding success. Green and Giese believed that "Short-term effects can cumulatively develop into long-term impacts, and effects on individual animals can cumulatively affect populations and ecosystems" (p. 81). At the ecosystem level, sport tourism and recreation impacts influence species that people are less aware of. Therefore, the presence of sport tourism and recreation users can change the balance of ecosystems.

Environmental Impacts of Materials Used in Sport Tourism and Recreation

Although not heavily emphasized in the literature, materials used in sport tourism and recreation can have a considerable impact on the environment. In recent decades, sport tourism and recreation activities are highly dependent on technology and sporting equipment. Many new designs and materials are manufactured to satisfy the needs of athletes and users, but most of these materials have shorter life cycles compared to previous generations (Subie et al., 2009). Therefore, the environmental impacts of sport equipment have increased over time. In addition, sport clothing and equipment have become fashionable in recent decades. Fashion trends in sport and recreation motivates people to purchase more sport tourism and recreation-related materials, posing another environmental challenge related to sport tourism and recreation (Joy et al., 2012).

Environmental Impacts of Decentralized Sport: Commuting and Transportation

Another issue with the modern approaches in recreation and tourism is transporta- tion (Bunds et al., 2018). Most outdoor sport tourism and recreation opportunities require transportation. As public transportation is not an option for many sites, peo- ple must commute using personal vehicles. These decentralized sites are a cause for billions of trips to sport tourism and recreation facilities every year that result in significant environmental impacts.

Positive Impacts

Although negative environmental impacts of sport tourism and recreation have been emphasized in the literature, several positive impacts have also been high- lighted. These positive impacts are elaborated below (Mason, 2015).

Conservation efforts. Ecosystem components are often the main attractions for sport tourism and recreation activities. For example, sport hunting, wildlife view- ing, and many other sport tourism and recreation activities are solely dependent on wildlife pretense. In many countries, people who are interested in these activities contribute generously to the conservation of ecosystems. For instance, in Canada, hunters significantly contribute to the conservation of species through donations and nest box set-up and maintenance.

Protected area designation. Many sport tourism and recreation activities hap- pen outdoors. This encourages planners and managers to establish protected areas in order to serve recreation users while conserving natural resources. Also, the revenue generated through these activities goes toward the protection of these areas as well as endangered species.

Investments in infra-structure and super-structure. To provide access for the users, planners need to invest in the infra-structures and super-structures. These facilitate the conservation of ecosystems by providing waste management services and better road access.

Crowd-sourced services. People who are engaged in sport tourism and recrea- tion activities often develop a sense of place for the areas that host their activities. Research results have shown that place attachment encourages people to partake in activities, such as volunteering to clean natural environments and collecting in- vasive species. These activities are often organized by NGPs who are interested in different sport tourism and recreation activities.

Land use change. High demand of sport tourism and recreation activities in some areas makes it possible to revive abandoned areas or to change the use of unproduc- tive lands. These activities also enable people to utilize vacant/unused buildings.

Impact Management

This section briefly reviews the most popular impact management tools and approaches used in sport tourism and recreation management. First, Limits of Acceptable Change (LAC) will be introduced. Then, carrying capacity analysis, one of the most popular user management techniques, is discussed. These two methods utilize threshold-based techniques to determine the highest number of users a site can host without causing excessive impacts. Linear and non-linear thresholds approach as well as concentrated and dispersed uses are explained.

Limits of Acceptable Change (LAC)

As a management process, LAC introduces nine steps that planners can take to manage the increasing number of users. It does so by comparing the existing situation with the desired conditions (Ahn et al., 2002; Bentz et al., 2016; McCool, 2013; Stankey et al., 1985). In the first step, stakeholders (i.e., communities, user groups, and managers) identify and define the issues and concerns associated with the resource. In the second step, planners define different social, natural resource, or managerial opportunity classes and zones (condition classes). The goal of this step is to prepare a range of conditions that align with the existing situation. In the third step, planners develop a series of indicators to be able to measure the conditions in each zone. These indicators are often created by the help of users through rating the importance of various conditions. The fourth step focuses on the inventory of the existing conditions based on the indicators. Results of these inventories provide baseline data that guide the standard developments. The inventory of the conditions is often conducted in the field by using multiple measurement tools such as impact assessment tools and questionnaires. In step five, planners identify measurable standards for each opportunity class. This step enables managers to quantify standards, specify aspects of the indicators, and express them in terms of probability. Step six is focused on the identification and allocation of alternative opportunities. Essentially, it analyses the data collected based on the indicators. Products of this step are usually tables and maps as well as summaries and zone allocations. Step seven compares the existing conditions with the defined standard conditions. This step produces outputs that show where the problems exist. For example, it provides maps that reflect impacted areas or areas that do not meet the defined standards. Step eight selects the alternatives or the management program that is focused on the problem identified in the previous step. The final step implements the management program and monitors the changes. A feedback system is implemented in this step to guide the progress and suggest future changes. LAC is a process that never ends (see Figure 4.2). The monitoring system in step nine guides the development of the second round of assessment, analysis, and implementation. Next, we will discuss how LAC studies have identified several gaps that led researchers to design and develop a visitor use management system.

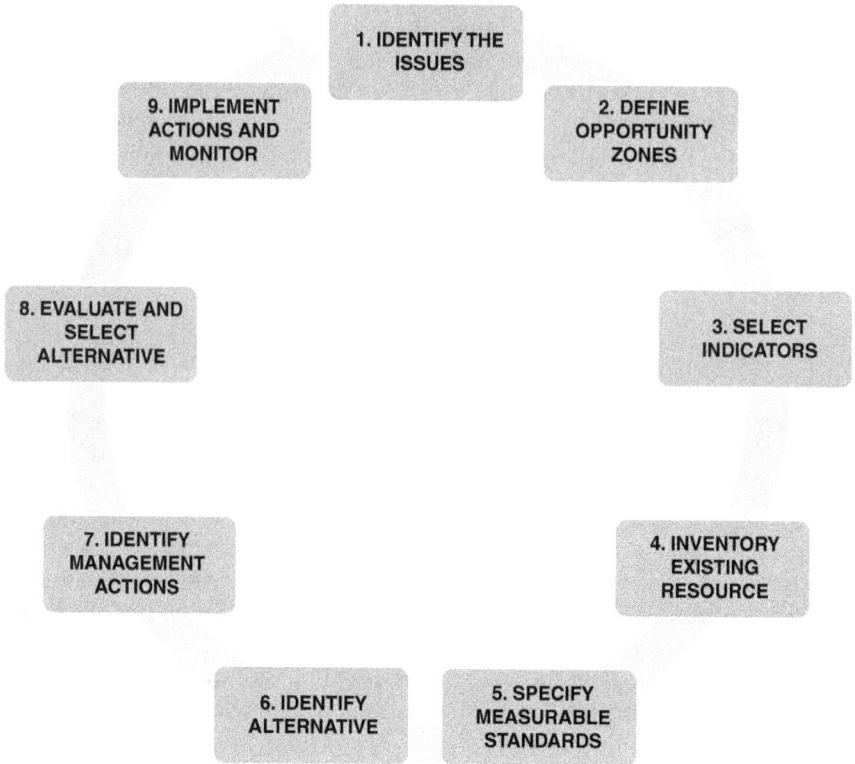

Figure 4.2 LAC Process.
Source: Stankey et al. (1985).

User Management Framework

As mentioned, LAC has been used as a tool for planners to manage the impacts of different sport tourism and recreation activities. Several case studies led planners to develop user management frameworks, such as carrying capacity analysis. These techniques have been revised, improved, and implemented in multiple sport tourism and recreation destinations around the world. Interagency Visitor Use Management Council's visitor use management framework (Marion, 2016) defines four elements and 14 steps (Figure 4.3).

Element 1: Build the foundation. This element explains how to best approach the project, organize the tasks, check availability of resources, and plan the project. The four steps of this element include: (1) clarification of the project purpose and needs; (2) review of the area's purpose(s) and applicable legislation, agency policies, and other management directions; (3) assessment and summarization of the existing information and current conditions; and (4) development of a project action plan. Based on the data obtained in the previous steps, the action plan and timeline of the project are determined.

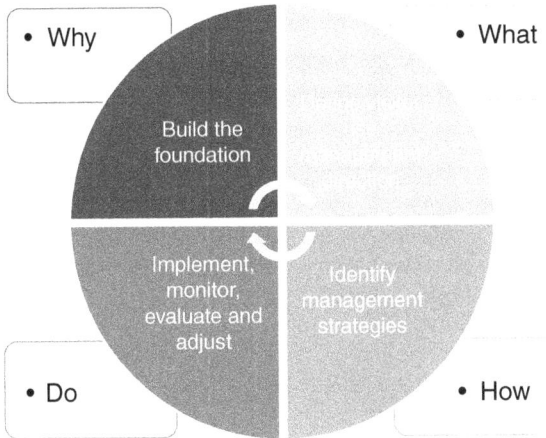

Figure 4.3 Visitor Management Framework.

Source: VUMF (2016).

Element 2: Define visitor use management direction. This element focuses on monitoring planning approaches to future changes and trends. The three steps of this element include: (5) defining the desired conditions; (6) defining appropriate visitor activities, facilities, and services; and (7) selection of the indicators and threshold establishment.

Element 3: Identify management strategies. This element includes management's decision to create and identify strategies as solutions to the known problems. This element includes steps eight to eleven: (8) comparing the differences between existing and desired conditions; (9) identifying visitor use management strategies; (10) identifying visitor capacities and strategies to manage use level; and (11) developing a monitoring strategy.

Element 4: Implement, monitor, evaluate, and adjust. This element is a feedback process that provides insight on the effectiveness of the strategies. The insight gathered will guide the next stages of the management plan. Steps of this element focus on: (12) implementation of management actions; (13) conducting an ongoing monitoring and evaluation process; and (14) adjusting management actions in order to achieve desired conditions.

Conclusion

The aim of this chapter was to introduce different impacts of sport tourism and recreation. After defining environments in which sport tourism and recreation occur, this chapter expanded on the different positive and negative impacts that sport tourism and recreation activities have on the environment. Multiple classifications of impacts were introduced and discussed. Finally, a few of the most effective impact management approaches were discussed. It is worth mentioning that environmental, economic, social, and cultural impacts of sport tourism and recreation go hand-in-hand, each influencing the other. Therefore, all four dimensions need to be considered in the management of sport tourism and recreation destinations.

References

Ahn, B., Lee, B., & Shafer, C. S. (2002). Operationalizing sustainability in regional tourism planning: An application of the limits of acceptable change framework. *Tourism Management, 23*(1), 1–15.

Bentz, J., Lopes, F., Calado, H., & Dearden, P. (2016). Sustaining marine wildlife tourism through linking limits of acceptable change and zoning in the Wildlife Tourism Model. *Marine Policy, 68*, 100–107.

Buckley, R. (2004). *Environmental impacts of ecotourism.* Cambridge.

Buckley, R. (2005). Recreation ecology research effort: An international comparison. *Tourism Recreation Research, 30*(1), 99–101.

Buckley, R. (2011). Tourism and environment. *Annual Review of Environment and Resources, 36*, 397–416.

Bunds, K. S., Kanters, M. A., Venditti, R. A., Rajagopalan, N., Casper, J. M., & Carlton, T. A. (2018). Organized youth sports and commuting behavior: The environmental impact of decentralized community sport facilities. *Transportation Research Part D: Transport and Environment, 65*, 387–395. https://doi.org/10.1016/j.trd.2018.08.017.

Cole, D. N. (1989). Recreation ecology: What we know, what geographers can contribute? *The Professional Geographer, 41*(2), 143–148.

Cole, D. N. (2004). Impacts of hiking and camping on soils and vegetation: A review. *Environmental Impacts of Ecotourism, 41*, 60.

Dunlap, R. E. (2008). The new environmental paradigm scale: From marginality to worldwide use. *The Journal of Environmental Education, 40*(1), 3–18.

Fox, J. (2019). *A tourism impact index for water-based natural attractions field-tested in subarctic and maritime climates* [Master's Thesis, Western Kentucky University]. https://digitalcommons.wku.edu/cgi/viewcontent.cgi?article=4145&context=theses

Godtman Kling, K., Fredman, P., & Wall-Reinius, S. (2017). Trails for tourism and outdoor recreation: A systematic literature review. *Turizam: međunarodni znanstveno-stručni časopis, 65*(4), 488–508.

Green, R., & Giese, M. (2004). Negative effects of wildlife tourism on wildlife. In K. Higginbottom (Ed.), *Wildlife Tourism: Impacts, Management and Planning* (pp. 81–97). Common Ground Publishing Pty Ltd. Alton, Australia.Hammitt, W. E., & Cole, D. N. (1998). *Wildland recreation: Ecology and management* (2nd Ed.). John Wiley & Sons.

Hammitt, W. E., Cole, D. N., & Monz, C. A. (2015). *Wildland recreation: Ecology and management.* John Wiley & Sons.

Joy, A., Sherry, Jr., J. F., Venkatesh, A., Wang, J., & Chan, R. (2012). Fast fashion, sustainability, and the ethical appeal of luxury brands. *Fashion Theory, 16*(3), 273–295.

Lakhouit, A. (2019). Tourism impact on marine ecosystems in the north of Red Sea. *Journal of Sustainable Development, 13*(1), 10. https://doi.org/10.5539/jsd.v13n1p10

Leave No Trace Center for Outdoor Ethics. (2020). *The 7 principles.* Retrieved from https://lnt.org/why/7-principles/

Li, G. (2008). The nature of leisure travel demand. In A. Graham, A. Papatheodorou & P. Forsyth (Eds.), *Aviation and tourism* (pp. 37–50). Routledge. https://doi.org/10.4324/9781315568522-12

Liddle, M. J. (1991). Recreation ecology: Effects of trampling on plants and corals. *Trends in Ecology & Evolution, 6*(1), 13–17.

Magro-Lindenkamp, T. C., & Leung, Y. F. (2019). Managing environmental impacts of tourism. In S. F. McCool & K. Bosak (Eds.), *A research agenda for sustainable tourism* (pp. 223–238). Edward Elgar Publishing.

Marion, J. L. (2016). A review and synthesis of recreation ecology research supporting carrying capacity and visitor use management decision-making. *Journal of Forestry, 114*(3), 339–351.

Marion, J. L., Leung, Y. F., Eagleston, H., & Burroughs, K. (2016). A review and synthesis of recreation ecology research findings on visitor impacts to wilderness and protected natural areas. *Journal of Forestry, 114*(3), 352–362.

Mason, P. (2015). *Tourism impacts, planning and management* (3rd Ed.). Taylor and Francis.

McCool, S. F. (2013). Limits of acceptable change and tourism. In A. Holden & D. A. Fennel (Eds.), *Routledge handbook of tourism and the environment* (pp. 285–298). Routledge.

Monz, C. A., Pickering, C. M., & Hadwen, W. L. (2013). Recent advances in recreation ecology and the implications of different relationships between recreation use and ecological impacts. *Frontiers in Ecology and the Environment, 11*(8), 441–446.

Mosisch, T. D., & Arthington, A. H. (2004). Impacts of recreational power-boating on freshwater ecosystems. In R. Buckley (Ed.), *Environmental impacts of ecotourism, ecotourism book series, volume 2* (pp. 125–154). CABI Publishing. https://doi.org/10.1079/9780851998107.0125

Petrosillo, I., Valente, D., Pasimeni, M. R., Aretano, R., Semeraro, T., & Zurlini, G. (2019). Can a golf course support biodiversity and ecosystem services? The landscape context matter. *Landscape Ecology, 34*, 2213–2228.

Pickering, C. M., & Hill, W. (2007). Impacts of recreation and tourism on plant biodiversity and vegetation in protected areas in Australia. *Journal of Environmental Management, 85*(4), 791–800.

Stankey, G. H., Cole, D. N., Lucas, R. C., Petersen, M. E., & Frissell, S. S. (1985). *The limits of acceptable change (LAC) system for wilderness planning.* (General Technical Report INT-176). United States Department of Agriculture. Retrieved April 10, 2024 from https://www.fs.usda.gov/Internet/FSE_DOCUMENTS/stelprdb5346594.pdf

Stefanic, J., & Delgado, R. (1996). *No mercy: How conservative think tanks and foundations changed America's social agenda.* Temple University Press.

Subie, A., Mouritz, A., & Troynikov, O. (2009). Sustainable design and environmental impact of materials in sports products. *Sports Technology, 2*(3–4), 67–79.

Sun, D., & Walsh, D. (1998). Review of studies on environmental impacts of recreation and tourism in Australia. *Journal of Environmental Management, 53*(4), 323–338.

Swarbrooke, J. (1999). *Sustainable tourism management.* CABI Publications.

VUMF. (2016). Visitor Use Management Framework. *A guide to providing sustainable outdoor recreation.* Retrieved from https://visitorusemanagement.nps.gov/VUM/Framework

Wagar, J. A. (1964). The carrying capacity of wild lands for recreation. *Forest Science, 10*(suppl_2), a0001-24.

5   Crowding in Outdoor Sports, Recreation, and Tourism

A Review of American and Canadian Parks and Protected Areas

Hannah R. Dudney and Farhad Moghimehfar

What's Happening in Parks?

Parks and protected areas (PPAs) are enticing spaces for outdoor physical activities used by tourists and residents for relaxation, escape, peace, tranquility, health, and more. However, what happens when this expected environment no longer exists or is too busy? Due to several reasons, including promoting nature-based sports and recreation, improved accessibility, and increased awareness of the benefits of outdoor sports, parks and protected areas have become overwhelmingly busy. For instance, in the lower mainland and coastal mountains of British Columbia in Canada, park planners have faced a nearly 400% increase in park visitors in the past three years (D. Smith, personal communication, November 2021). Such an increase is evident in many parks and protected areas in North America. Tourists and locals visit parks for numerous sports and recreation, such as hiking, camping, museums, and various aesthetic, social, and psychological benefits (Bhalla et al., 2021; Chikuta & Saayman, 2017; Winter et al., 2019). In addition to these motivators, nature-based physical activities have become increasingly popular and important due to numerous social problems, such as urbanization, mental health, work patterns (Elmahdy et al., 2017), and in recent years, coping with the COVID-19 pandemic impacts (Bhalla et al., 2021). This chapter reviews the literature on parks and crowding in North America. First, we define the context, parks, and protected areas. Then we expand on the significance of studying and managing crowding in parks and protected areas. Finally, we discuss different management approaches and review case studies.

Although the discussion in this chapter will focus mainly on national and provincial parks, we recognize that this review of the literature could be applicable and relevant to other popular nature-based tourist destinations. To define parks and protected areas more concretely, many attempted to focus on a specific park type. Parks can fall under the definition of a Class A state and provincial park (e.g., Garibaldi Provincial Park, BC, Canada) defined as "lands dedicated to the preservation of their natural environments for the inspiration, use and enjoyment of the public" (BC Parks, n.d.c, p. 8). National Parks (e.g., Banff National Park, AB, Canada.) are also often defined as natural areas protected for education, enjoyment, and sustainable maintenance (Parks Canada, 2019). Other

DOI: 10.4324/9781003476658-7

popular green spaces considered protected areas might be categorized as peri-urban parks. Peri-urban parks are "nature-based attractions located on the out-skirts of cities" (Zhang et al., 2021). With various definitions of parks and pro-tected natural spaces, it is important to consider all definitions under the umbrella of nature-based recreation destinations.

The popularity of such nature-based destinations has led to increased national and provincial park activity in North America, especially during peak tourist sea-sons. This leads to over-visitation and crowding, threatening the quality of visitor experiences for all (Bergstrom et al., 2020; Dodds & Butler, 2019; Kohlhardt et al., 2018; Manning & Anderson, 2012). However, crowding in popular PPAs is not a new concept. Thus, scholars in North America and beyond have explored ways in which to manage crowding with the hope of achieving the dual mandate of PPAs, to preserve the natural environment while also ensuring public access and enjoy-ment (Manning, 2002; National Park Service, 2015; Parks Canada, 2002; Weber et al., 2019).

In a broader context, tourism crowding, also known as over-tourism (Redko et al., 2022), is well studied in academic literature in a wide range of tourist des-tinations. Over-tourism has the ability to not only disrupt the visitor experience but especially disrupt the daily lives of local residents (Dodds & Butler, 2019). In discussions, such as Dodds and Butler's (2019) review of over-tourism, the great debate of when to restrict visitation and how to do it is a common concept for tour-ist destinations, no matter their genre. Many destination managers have the sense that something must be done, but it seems as though every option to manage the ad-verse effects of over-tourism will be met with equal amounts of scrutiny from visi-tors, residents, or local authorities (Dodds & Butler, 2019). Expanding on Dodds and Butlers' points about urban over-tourism, a more significant concern now be-comes how to manage nature-based tourism destinations like PPAs while balanc-ing the need for jobs, public enjoyment, and tourist-driven revenue alongside the need to avoid overwhelming the tourists, locals, and businesses. Dodds and Butler (2019) state that widespread acknowledgment of over-tourism must happen before essential changes occur.

Significance of Crowding in Parks and Protected Areas

In contrast to crowding in built environments (e.g., shopping malls or amusement parks), crowding in parks and protected areas poses a specific and more complex problem. Parks can be easily accessible ways for people in an increasingly urban world to enjoy nature (Maller et al., 2009). Thus, it is logical that parks and re-lated outdoor and nature-based leisure are widely recognized proponents of physi-cal, social, and psychological health. In a quantitative study on nature tourism in parks, Buckley (2020) found that visiting parks significantly influenced health and happiness, supporting other studies that recognize positive relationships between park visitation and visitor well-being (Maller et al., 2009; Romagosa et al., 2015). The benefits of visiting parks are identified by a widely known movement called Healthy Parks, Healthy People (Maller et al., 2009).

The Healthy Parks, Healthy People movement is a global initiative that draws attention to parks as more than just natural resources for preservation but as sources of well-being for visitors (Bricker et al., 2016). Maller et al. (2009) summarize park visitation benefits, including physical, mental, spiritual, and environmental, highlighting that parks make ideal spaces to address a wide range of factors that affect well-being. Maller et al. compile suggestions for park management agencies to help people re-envision parks and recognize their importance: share the essential benefits of parks with the general community, educate governments about the importance and use of parks, and facilitate community engagement with nature.

It makes sense that if parks can improve health and well-being, lack of parks (Maller et al., 2009) or access to "unhealthy" parks may have the opposite effect, and who for? In fact, Maller et al. explored studies that suggest keeping humans from interacting with nature may have detrimental effects on their energy levels (Katcher & Beck, 1987; Stilgoe, 2001) and keep them from addressing human needs for exercise and touch (Katcher & Beck, 1987). More recently, scholars such as Mark Groulx et al. (2021) made the link between accessible nature spaces for people with disabilities and the power of equal participation in parks for all to live healthy lives. These considerations become even more critical when the park environment is impacted by over-tourism.

Outdoor sports crowding in North America is not a new concept or concern for park management (Shelby et al., 1989; Wagar, 1964; Wall, 2020). Popular parks in Canada and the United States of America (U.S.) are no strangers to crowded parking lots, streets, trails, facilities, and even surrounding towns, especially during high tourist seasons. For example, parks such as Banff (Ross, 2021) and Jasper National Park (Parks Canada, 2021) in Alberta, and Pacific Rim National Park (Robinson, 2020), Joeffre Lakes Provincial Park (CBC News, 2021) in British Columbia (BC) experience problems with seasonal crowding. In the United States, parks such as Yosemite National Park (National Park Service, 2022b) and Sequoia and Kings Canyon National Park (National Park Service, 2022a) in California, Arches and Canyonlands National Parks in Utah (Siegler, 2021), and Yellowstone National Park in Wyoming (USA Today, 2021) are just a few of many parks that report troublesome crowding. All these parks have been struggling with managing overcrowding to improve the visitor experience and protect the integrity of the natural habitats. Several approaches to managing crowding have been invented or implemented in these parks. The next section of this chapter highlights these methods.

Approaches to Managing Crowding in Parks and Protected Areas

Dodds and Butler (2019) suggested that many solutions to over-tourism are band-aid solutions that fail to address the root of the problem, often failing to fix the problem or even making it worse. Solutions implemented by tourist destinations have commonly been to convince visitors to visit less crowded destinations instead (Dodds & Butler, 2019). On the contrary, these types of solutions proved to be troublesome. For example, Butler (2001) discovered that when a destination attempts to market their destination for the off (or less popular) season, it increased year-round visitation and had little effect on peak (busy) season visitation. This is especially true when the locations are exceptionally well known, such as the Pyramids in Cairo, Egypt (Dodds & Butler, 2019).

More specifically, in PPAs, planning and management teams face with a more complex problem of over-tourism. Thus, scholars have attempted to devise scientific, researched-based methods to guide management tactics, including carrying capacity, indicators and standards of quality, and limits of acceptable change, many of which lend to management frameworks such as the Visitor Use Management Framework (VUMF; US National Parks Services https://visitorusemanagement.nps.gov/).

Physical Carrying Capacity

From a management perspective, crowding is often conceptualized as the point when park facilities and spaces reach capacity (Krinsky & Kuehn, 2020). In other words, when the amount of people present begins to harm the natural environment (Manning, 2002). Many solutions to park crowding involve finding an exact number of people a space can hold before it damages the physical environment (e.g., soil compaction). As stated by Manning (2002), the concept of carrying capacity in the parks realm first only focused on nature preservation, but later, social aspects of visitor capacity were explored.

It was not until scholar J. Alan Wagar (1964) introduced the notion that we must look broader and consider human values that the shift toward more than physical carrying capacity began (Manning, 2002). Today, it is common for scholars to seek out physical and environmental carrying capacity while also considering other diverse factors outside of the natural environment. For example, Wiyono et al. (2018) in their study of the visitor carrying capacity of Peucang Island measured the length and width of the track, land slope, and soil type. They developed an equation for physical carrying capacity that pre-determined the area required per user to walk comfortably. Their calculated area in their study equaled one squared meter per visitor, along with the trail distance and number of visits per day. Suana et al. (2020) investigated the environmental carrying capacity of bird-watching in Kerandangan Natural Park. Similarly, their analysis used the "width and length of the trails, the visit time, as well as the soil texture and slope along the trails" (p. 2266). Although these more recent works calculate environmental carrying capacities, they recognize the need for a holistic approach to carrying capacity analysis.

Suana et al. (2020) and Wiyono et al. (2018) also recognize the need to find a Real Carrying Capacity, which is the "maximum number of visits after considering the factors that limit the occurrence of visits or correction factor (CF)" (Suana et al., 2020, p. 2268). The correction factor considers "physical, environmental, ecological, social, and management variables of the area" (Suana et al., 2020, p. 2268). These works include aspects of social carrying capacity, as opposed to strictly focusing on the physical impacts of increased park visitation. Additional scholarly works now focus strictly on how the increase in visitors affects the quality of park experiences, attempting to discover a social carrying capacity.

Social Carrying Capacity

The traditional definition of park crowding targets a point when an area reaches its physical or environmental carrying capacity regarding visitor numbers. The more recent approaches, however, see crowding as "the point when the number of users negatively impacts visitor experiences" (Krinsky & Kuehn, 2020, p. 128). Alternatively to physical carrying capacity, social carrying capacity recognizes that perceptions of crowding are "subjective and person-specific" (Gonson et al., 2018, p. 148). For this reason, Manning (2013) states that to honor the visitors in the search for carrying capacities, park visitors themselves should be included in the development of values or objectives. Social carrying capacity focuses on the quality of the user experience (Gonson et al., 2018), rather than how the number of visitors effects the environment or man-made infrastructure. Important factors to consider when surveying visitors for social carrying capacity included personal characteristics like previous visits to the area, socio-economic status, gender, city of origin, trip duration, group size, preferences about the location, and why they chose to visit (Gonson et al., 2018).

Surveys that are used to determine social carrying capacity are often attempting to discover person or group norms for park visitors (Manning, 2013). By surveying the acceptability of certain park features like crowding, scholars can graph responses to develop social norm curves (see Figure 5.1). This creates a visual tool for management to see the preferred park conditions, the range of acceptable conditions, and the maximum acceptable conditions. These normative curves will be used to advise standards of quality (Manning, 2013), as discussed next.

When carrying capacity is applied to humans, as is social carrying capacity, it must be less rigid than previous methods that claim to be scientific (Manning, 2013). Consequently, many scholars have come to criticize the concept of carrying

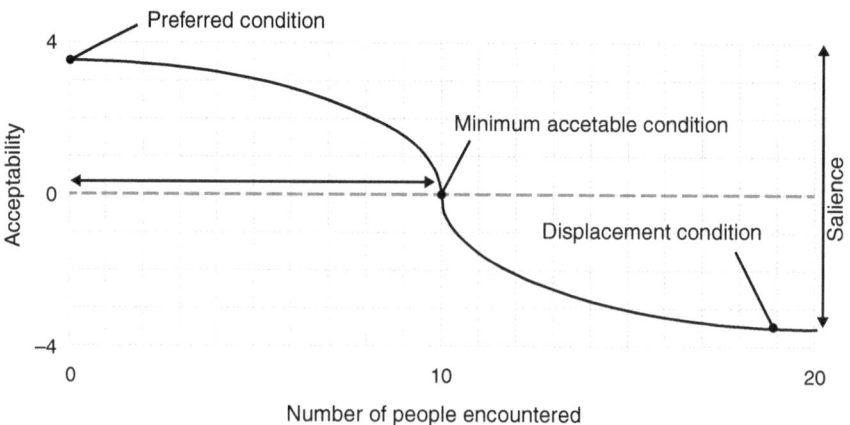

Figure 5.1 Sample Social Norm Curve.

capacity and its quest for a specific number. More recent works have highlighted that carrying capacity is far too simplified to manage problems related to visitation in parks (Marion, 2016). Manning (2002) also notes that carrying capacity was essentially a failure in a parks realm because, regardless, some negative human impacts were bound to happen, making it almost pointless (and arguably impossible) to quantify at which point this happens. For that reason, Manning suggests that the key is to discover how much damage (environmental) or discomfort (social) due to the presence of people is acceptable. This can be discovered by exploring visitors' indicators and standards of quality in parks.

Indicators and Standards of Quality

Indicators and standards of quality lay the groundwork for many park management frameworks, such as the Limits of Acceptable Change (McCool, 2013), discussed next. The goal of discovering indicators and standards has been a long-standing concept in the field of parks and outdoor recreation (Manning, 2011). Indicators are specific, measurable features that management teams use as checkpoints to see if they have met management objectives, while standards of quality mark the minimum requirements of the management objectives (Manning & Anderson, 2012). Manning (2011) suggests that these specific indicators can be discovered through communication with various park stakeholders (e.g., interviews) to discuss aspects of the visitor experience. For example, Manning (2011) conducted interviews and focus groups with park visitors to the popular Arches National Park in Utah, U.S., to discover indicators related to crowding. Some indicators included crowded attractions and soil damage. These indicators along with others were used as reference points to devise standards of quality.

When indicators and standards of quality note the expectations of visitors, managers can also move forward with other management frameworks. For example, they can discover where parks exist on the Recreation Opportunity Spectrum, which, in its simplified version (Manning & Anderson, 2012), measures park opportunities from wild to urban in categories of resource conditions, experiential conditions, and managerial conditions. Urban parks having mainly domesticated animals present, low levels of solitude, and high levels of development, while wild parks would have wild animals present, high levels of solitude, and no development. By discovering where a park exists on this spectrum, park management will know what standards to maintain in their goals and can inform the use of other frameworks like the Limits of Acceptable Change (LAC) framework.

Limits of Acceptable Change

One way to address crowding is to monitor LAC. LAC is a concept that has not only applied to crowding-related park variables but, in general, has been a guiding format for park management. In the case of crowding, discovering visitors' limits of acceptable change will hope to answer the question, "what level of perceived crowding should be allowed?" (Manning & Anderson, 2012, p. 6). This framework assumes that any amount of human disruption to an environment will alter the natural and social environment and that management strategies are necessary components to control the impact of human presence (McCool, 2013). For this reason, the LAC framework is frequently used in nature-based tourism context in general, including parks (McCool, 2013). McCool states that the LAC framework was created largely in response to the failures of carrying capacity measurements. It views these limits only as guidelines, not as fixed numerical values that must be obtained (Zelenka & Kacetl, 2014).

Threefold Framework of Outdoor Recreation

Related to these previous management techniques to address issues like crowding, Manning and Anderson (2012) discuss the importance of the Threefold Framework of Outdoor Recreation. The threefold framework recognizes environmental (resources), social (experiences), and managerial (regulations) as the three important components of outdoor recreation like parks. They also note that these components interact with one another. For example, crowding can be caused or allowed by management decisions which can damage natural resources and social experiences in parks. Alternatively, management tools and techniques such as raised, wide boardwalks over fragile vegetation, such as that in BC's Chun T'oh Whudujut Provincial Park (Groulx et al., 2021), can benefit the park ecosystem while improving the social experiences of many visitors.

Using indicators and quality standards, the LAC framework, physical carrying capacity, and social carrying capacity aim to provide structure and frameworks for management teams to continually discover, address, and assess visitor and environmental needs. For example, more comprehensive visitor use frameworks like the VUMF (Visitor Use Management Framework) consider these strategies and are used by large parks organizations like the U.S. National Park Service (Interagency Visitor Use Management Council, n.d.) and Canada's BC Parks uses similar strategies by reviewing park values and including visitor feedback in their management plan strategy (BC Parks, n.d.b).

Scholars such as McCool and Lime (2001) argue that frameworks like LAC, as well as carrying capacities, do not allow managers to resolve these complex problems. However, they still serve as platforms to attempt to manage these problems in parks (McCool & Lime). This is reflected in the Interagency Visitor Use Management Council's more recent guide, specifically combining concepts of the VUMF and park capacity (Interagency Visitor Use Management Council, 2019). This guidebook assists park management in understanding, identifying, and managing capacity in PPAs and similar spaces. These frameworks also highlight the benefits of involving a variety of park stakeholders, not just management, in the process of managing public lands (McCool & Lime, 2001).

Applied strategies such as the Visitor Capacity Guidebook (Interagency Visitor Use Management Council, 2019) are often used in a park-specific or even attraction-specific manner to devise plans that specifically suit those park features. Thus, I will elaborate on how some of the strategies mentioned previously have been used in real-world scenarios, exploring two case studies, one in Arches National Park in Utah, U.S., and another in Garibaldi Provincial Park in British Columbia (BC), Canada.

Practical Implications: Studies of Over-Visitation

As the rise of park visitation and subsequent consequences become more of a concern, many scholars continue to explore how the amount of park visitors does or will affect the experiences of park stakeholders (e.g., tourists, park managers, park staff, local visitors, residents). Thus, in this chapter, we review four cases, two in the U.S. and two in Canada, that explore the concern of crowding in PPAs. After a summary of each case, we will further review their relation to one another. This review aims to analyze how different problems related to crowding must be approached in different ways, as well as how different approaches will lead to different suggestions for park management.

Crowding in Arches National Park, Utah, U.S.

In Southeast Utah, U.S., Arches National Park (NP) is home to unique desert sandstone formations that have made it a must-see attraction in the U.S. and home to one of the most popular trails in all of the U.S. national parks (Manning & Anderson, 2012). Manning and Anderson report 1 million yearly visitors to Arches NP in their 2012 text. More recently, the U.S. National Park Service (NPS) reports that over 1.8 million visitors entered Arches NP in 2021 (National Park Service, n.d.). For this case, park scholars Robert Manning and Laura Anderson (2012) focused on the social consequences of this rise in visitation, as opposed to its ecological impacts.

To address this problem, Manning and Anderson (2012) first zoned the park into sections, labeling them based on their level of development or how primitive they are. Next, they discovered the limits of acceptable change. To discover social limits of acceptable change, management first discussed proposed limits and then surveyed visitors to further explore these limits. Additionally, they proposed to discover a group of social norm curve (as mentioned previously) for the limitations at a popular attraction in Arches NP. To discover the social norms' curve for crowding, Manning and colleagues developed images that represented different levels of crowding at the Delicate Arch attraction. They then asked visitors in that area to decide how acceptable each of the scenarios was.

Using these findings, Manning and colleagues were able to quantify the level of acceptance for crowding at this specific attraction. These findings were used to plan the number of spots in subsequent parking lots, based on how parking spot availability influenced the number of people at these attractions. Another example provided by Manning and Anderson (2012) was to use these findings to create a limit for park reservation systems in order to limit the number of people who can enter a park on a given day. Statistical models were used to monitor these relationships over time.

COVID-19 and Outdoor Recreation Crowding

During the ongoing COVID-19 pandemic, William Rice et al. (2022) decided to take an alternative look at crowding perceptions in outdoor recreation spaces. Rice et al. surveyed outdoor recreation users in the U.S. to begin to explore how social norms of outdoor recreation crowding have shifted since the COVID-19 pandemic began. Due to the increased use of face masks, often as a government mandate or personal choice, Rice et al. predicted that the use of face masks in outdoor recreation would change how people perceive crowds.

Rice et al. (2022) conducted an online survey of mostly U.S. outdoor enthusiasts, purposively selected using email distribution to members of the Leave No Trace Center for Outdoor Ethics to explore these new perceptions. The survey had visual components corresponding with acceptability ratings that allowed Rice et al. to conduct a normative assessment regarding visitor density, crowding, and visitor capacity. Using both amount of people (1, 3, 6, 12, 18) and amount of mask-wearing people (all masked, some masked, none masked) as indicators. Rice et al. compared acceptability across various situations. Additionally, they ensured that people in the photos were ethnically diverse to simulate the actual visitor population (informed by Stanfield et al., 2005 and Xiao et al., 2020).

Rice et al. (2022) discovered that collective mask-wearing was a strong indicator of acceptability, meaning that the more people wearing masks, the more acceptable crowding conditions were, no matter how many people were present. The findings led Rice et al. to suggest norm-inducing signage in outdoor recreation spaces, considering it will improve the park experience, with crowds or without.

Resident Perceptions of Tourism in Banff National Park, AB, Canada

Resident perceptions of increasing tourism density remain an important perspective for park planners to consider. Thus, scholars like Fangbing Hu et al. (2022) explored how residents' lives were affected by the increase of visitors to Banff National Park in Alberta, Canada. Being one of the world's most popular tourist destinations, Hu et al. (2022) decided to survey residents of Banff as well as residents in other nearby communities (Canmore and Golden). Although Hu et al. do not specifically focus on crowding perceptions, they discuss important aftereffects of increased park visitation, the need for communities to adjust infrastructure and ways of life when the number of visitors increases, commonplace responses to crowding or over-tourism.

To explore resident perceptions of tourism, Hu et al. (2022) used questionnaires that mainly focused on measuring residents' level of agreement or disagreement about tourism development in Banff National Park. Alongside the questionnaire,

Hu et al. compared the results to data from Statistics Canada to learn more about the communities' economic reliance and relationships with tourism (e.g., labor productivity and income from tourism).

Questionnaire responses reported signs of tourism crowding in the communities, as Hu et al. (2022) refer to as contamination issues. These issues consisted of noise, air, water, and light pollution. Banff residents, being considered a core community to the park's tourism, reported the most concern for noise pollution, lack of infrastructure, and lack of visitor environmental education programs. In combining findings from the questionnaire and the Statistics Canada data, Hu et al. found that Banff residents were estimated to be the most supportive of tourism development. They believe this could be related to their large population of tourism industry employees, leading us to consider the pros and cons of over-tourism. General suggestions for the community of Banff are to manage the number of visitors, the traffic flow and to promote environmental education to tourists.

Crowding on the Trails of Garibaldi Provincial Park, BC, Canada

Regan Kohlhardt et al. (2018) from the University of British Columbia and Simon Fraser University in BC, Canada, analyzed crowding in Garibaldi Provincial Park (PP). With the park's close proximity to the major city of Vancouver, BC, major highways, and scenic mountainous terrain, it continues to be a popular spot to visit (Kohlhardt et al., 2018). According to BC Parks (n.d.a), Garibaldi PP's number of visitors increased overall from 2007 to 2018, with the most recent collected year (2017/2018) reporting 137,885 total visitors.

To explore crowding, Kohlhardt et al. (2018) implemented a survey-based discrete choice experiment that examined how people made decisions based on different levels of crowding as well as different management approaches. Surveys were completed at select, busy trailheads in the park. They used digitally created photos focusing on six attributes in their scenarios, each with separate categories. Attributes included four which were communicated on the survey using images: viewpoint crowding (e.g., 24 people per 100 m), trail conditions (e.g., extreme erosion), trail crowding (e.g., 24 people per 100 m), and viewpoint conditions (e.g., a non-worthy attraction with less attractive views). Two additional attributes were communicated by text, the price of day use fees (e.g., $18), and driving distance to the nearest major city (e.g., 140 km from Vancouver).

Additionally, Kohlhardt et al. (2018) conducted a latent class analysis, which assists researchers in grouping respondents into categories based on specific attributes. For example, the outdoor tourist class was influenced the most negatively by crowding at the viewpoint and cared most about the worthiness (appearance) of the viewpoints. The goal of Kohlhardt et al.'s study was to support management in their pursuit to maintain visitor satisfaction while also upholding freedom of park access and ecological conservation. To achieve this, they provided detailed insight on the heterogeneity of park visitor needs for various topics related to crowding to thus guide management to more thoughtful decisions that satisfy a wide range of visitors.

Discussion

Crowding and over-tourism are at the core of these four cases, as the increase of visitors in these parks proves to affect the quality of park visitor experience (Kohlhardt et al., 2018; Manning & Anderson, 2012; Rice et al., 2022) and resident livelihood (Hu et al., 2022). Thus, scholars strive to suggest solutions to manage this visitation increase in hopes of striking a balance between nature conservation and park experience. In a more complex sense, sometimes it is not enough to look at nature conservation and park experience separately from diverse stakeholders, such as park residents (Hu et al., 2022) or separately from altering world events like the COVID-19 pandemic (Rice et al., 2022). This underlines the importance of re-adjusting, re-evaluating, and shifting approaches to park crowding, making studies of carrying capacity and crowding perceptions a field of outdoor recreation that must continue to grow and evolve.

Rice et al. (2022) highlighted that the COVID-19 pandemic has altered how tourists experience crowded spaces, even outdoors. Similarly, other scholars (e.g., Bhalla et al., 2021) note that nature-based tourism locations may continue to see a rise in visitors after the pandemic. In Arches NP, similar patterns seem to have resulted. According to the National Park Service (n.d.), Arches NP had a drop in visitors during the first year of COVID-19, 2020, and a rise in 2021. The year 2019 saw 1.6 million visitors, with a decline during the first year of the pandemic with 1.2 million in 2020, and a boom of visitation in 2021 with an all-time high of 1.8 million (National Park Service, n.d.). This increase is predicted by the COVID-driven positive sentiment toward nature-based travel destinations to "travel to natural and less crowded areas with a minimalist approach" (Bhalla et al., 2021, p. 776). Importantly, these COVID-driven park visitors, looking for less crowding and space away from others, may be met with disappointment when visiting popular parks. Cases such as Rice et al. (2022), which looked at the effect of mask-wearing on crowding perceptions in outdoor recreation spaces, highlight the importance of how changes to the outdoor environment (e.g., from a global pandemic) add an additional factor to consider in crowding management.

Considering the case from Manning and Anderson (2012) as a standard example of exploring crowding norms using visual discrete choice experiments, more recent scholars have also explored the addition of external factors. Similar to Rice et al.'s addition of mask-wearing as a new factor, Kohlhardt et al. (2018) also explored visitor choices based on crowding with the addition of distance from a major city, entrance fee, and natural conditions. These additional features help park staff gain a more complex sense of visitor expectations. Nonetheless, it proves important to also understand the visitors alongside their expectations.

Understanding various stakeholders in a more meaningful way is another strategy toward finding management strategies for crowding mitigation. Kohlhardt et al. (2018) study categorized respondents (park visitors) into distinct categories, so that park management could better understand the distinct expectations and desires of specific tourists regarding crowding as well as other park features like visitor fees. Rather than try to discover an exact number of capacity, Kohlhardt et al. (2018) recognized the importance of highlighting where and what variations of crowding should be managed as well as how different categories of visitors will be affected by crowding. This would allow management to strategize how to maintain economic well-being while satisfying visitor desires.

Hu et al. (2022) also recognized the importance of understanding stakeholders in a meaningful way. When studying resident perspectives on increased tourism in Banff National Park, they strictly included stakeholders that lived in the park communities, not visitors or management. This research will help avoid a single-minded approach to crowding management. By highlighting residents' concerns near a heavily visited park, Hu et al. (2022) did not investigate specific perceptions of crowding or try to discover a carrying capacity. Instead, they wanted to see what problems may have resulted from crowding in hopes management can address them more directly this way.

More recent studies related to crowding in PPAs, such as the ones mentioned in this section, highlight the continuing need for diverse perspectives, practical suggestions for park management, and the exploration of crowding in co-existence with cultural shifts.

Conclusion

Studies on crowding norms, limits of acceptable change, and carrying capacity in PPAs will support management to use funding strategically. Management decisions related to crowding and capacity are "best supported by understanding the desires and use patterns of stakeholders and collecting monitoring data" (Interagency Visitor Use Management Council, 2019, p. 19), including feedback from stakeholders other than management and staff. For example, park residents and a representative selection of visitors should also be included in capacity and crowding-related decisions and indicators. The VUMF states that crowding is subjective as well, and this is why comparing subjective accounts to the density of people in the space is a powerful way to measure experiences.

Along with the belief that crowding is subjective, scholars must ask themselves what stakeholders are included in measurements of limits of acceptable change and crowding norms. Are they visitors from out of town? Are they knowledgeable locals? Are they able to access online information prior to visiting? What are their demographic characteristics? How will these factors affect measurements of such a subjective experience within the realm of social sciences?

References

BC Parks. (n.d.a). *BC Parks 2017/18 statistics report*. Retrieved from https://bcparks.ca/research/statistic_report/statistic-report-2017-2018.pdf?v=1657316673397

BC Parks. (n.d.b). *Management planning process*. Retrieved from https://bcparks.ca/planning/process/

Bergstrom, J. C., Stowers, M., & Shonkwiler, J. S. (2020). What does the future hold for U.S. National Park visitation? Estimation and assessment of demand determinants and new projections. *Journal of Agricultural and Resource Economics, 45*(1), 38–55. https://doi.org/10.22004/AG.ECON.298433

Bhalla, R., Chowdhary, N., & Ranjan, A. (2021). Spiritual tourism for psychotherapeutic healing post COVID-19. *Journal of Travel & Tourism Marketing, 38*(8), 769–781. https://doi.org/10.1080/10548408.2021.1930630

Bricker, K. S., Brownlee, M. T. J., & Dustin, D. L. (2016). Healthy parks, healthy people. *Journal of Park and Recreation Administration, 34*(1), 1–3. Retrieved from https://go.exlibris.link/dl6436LN

Buckley, R. (2020). Nature tourism and mental health: Parks, happiness, and causation. *Journal of Sustainable Tourism, 28*(9), 1409–1424. https://doi.org/10.1080/09669582.2020.1742725

Butler, R. W. (2001). Seasonality in tourism: Issues and implications. In T. Baum & S. Lundtorp (Eds.), *Seasonality in tourism* (pp. 5–22). Elsevier.

CBC News. (2021). *B.C. calls for public feedback on crowd management plan at Joffre Lakes Provincial Park*. Retrieved from https://www.cbc.ca/news/canada/british-columbia/joffre-lakes-management-plan-1.5939790

Chikuta, O., & Saayman, M. (2017). Nature-based travel motivations for people with disabilities. *African Journal of Hospitality, Tourism and Leisure, 6*(1), 1–16. Retrieved from https://www.researchgate.net/publication/314276892_Nature_based_travel_motivations_for_people_with_disabilities

Dodds, R., & Butler, R. (2019). The phenomena of overtourism: A review. *International Journal of Tourism Cities, 5*(4), 519–528. https://doi.org/10.1108/IJTC-06-2019-0090

Elmahdy, Y. M., Haukeland, J. V., & Fredman, P. (2017). Tourism megatrends: A literature review focused on nature-based tourism. Norwegian University of Life Sciences. Report retrieved April 10, 2024, from https://hdl.handle.net/11250/2648159

Gonson, C., Pelletier, D., & Alban, F. (2018). Social carrying capacity assessment from questionnaire and counts survey: Insights for recreational settings management in coastal areas. *Marine Policy, 98*, 146–157. https://doi.org/10.1016/j.marpol.2018.08.016

Groulx, M., Lemieux, C., Freeman, S., Cameron, J., Wright, P. A., & Healy, T. (2021). Participatory planning for the future of accessible nature. *Local Environment, 26*(7), 808–824. https://doi.org/10.1080/13549839.2021.1933405

Hu, F., Wang, Z., Sheng, G., Lia, X., Chen, C., Geng, D., … Wang, G. (2022). Impacts of national park tourism sites: A perceptual analysis from residents of three spatial levels of local communities in Banff National Park. *Environment, Development and Sustainability, 24*(3), 3126–3145. https://doi.org/10.1007/s10668-021-01562-2

Interagency Visitor Use Management Council. (2019). *Visitor capacity guidebook, managing the amounts and types of visitor use*. Retrieved from https://visitorusemanagement.nps.gov/Content/documents/IVUMC_Visitor_Capacity_Guidebook_newFINAL_highres.pdf

Interagency Visitor Use Management Council. (n.d.). *Visitor use management framework*. Retrieved from https://visitorusemanagement.nps.gov/VUM/Framework

Katcher, A. H., & Beck, A. M. (1987). Health and caring for living things. *Anthrozoös, 1*(3), 175–183. https://doi.org/10.2752/089279388787058461

Kohlhardt, R., Honey-Rosés, J., Fernandez Lozada, S., Haider, W., & Stevens, M. (2018). Is this trail too crowded? A choice experiment to evaluate tradeoffs and preferences of park visitors in Garibaldi Park, British Columbia. *Journal of Environmental Planning and Management, 61*(1), 1–24. https://doi.org/10.1080/09640568.2017.1284047

Krinsky, A., & Kuehn, D. (2020). Managers' perceptions of crowding and noise in New York state parks. *The Journal of Park and Recreation Administration, 38*(4), 123–134. https://doi.org/10.18666/JPRA-2019-10080

Maller, C., Townsend, M., St. Leger, L., Henderson-Wilson, C., Pryor, A., Prosser, L., & Moore, M. (2009). Healthy parks, healthy people: The health benefits of contact with nature in a park context. *The GWS Journal of Parks, Protected Areas & Cultural Sites, 6*(2), 51–83.

Manning, R. E. (2002). How much is too much? Carrying capacity of national parks and protected areas. In A. Arnberger, C. Brandenburg, & A. Muhar (Eds.), *Monitoring and management of visitor flows in recreational and protected areas* (pp. 306–313). Boden-kultur University. Retrieved from http://npshistory.com/publications/social-science/how-much.pdf

Manning, R. E. (2011). Indicators and standards in parks and outdoor recreation. In M. Budruk & R. Phillips (Eds.), *Quality-of-life community indicators for parks, recreation and tourism management* (Vol. 43, pp. 11–22). Springer. https://doi.org/10.1007/978-90-481-9861-0_2

Manning, R. E. (2013). Social norms and reference points: Integrating sociology and ecology. *Environmental Conservation, 40*(4), 310–317. https://doi.org/10.1017/S0376892913000374

Manning, R. E., & Anderson, L. E. (2012). *Managing outdoor recreation: Case studies in the national parks.* CABI. Retrieved from https://ebookcentral.proquest.com/lib/viu/detail.action?docID=1044649

Marion, J. L. (2016). A review and synthesis of recreation ecology research supporting carrying capacity and visitor use management decisionmaking. *Journal of Forestry, 114*(3), 339–351. https://doi.org/10.5849/jof.15-062

McCool, S. F. (2013). Limits of acceptable change and tourism. In A. Holden & D. A. Fennel (Eds.), *Routledge handbook of tourism and the environment* (pp. 285–298). Routledge.

McCool, S. F., & Lime, D. W. (2001). Tourism carrying capacity: Tempting fantasy or useful reality? *Journal of Sustainable Tourism, 9*(5), 372–388. https://doi.org/10.1080/09669580108667409

National Park Service. (2015). *A call to action.* Retrieved from https://www.nps.gov/callto-action/pdf/c2a_2015.pdf

National Park Service. (2022a, May 16). *Sequoia & Kings Canyon traffic congestion.* Retrieved from https://www.nps.gov/seki/planyourvisit/traffic.htm#:~:text=Congestion%20on%20roads%20and%20in,by%20noon%2C%20and%20sometimes%20earlier.

National Park Service. (2022b, June 17). *Traffic in Yosemite National Park.* Retrieved from https://www.nps.gov/yose/planyourvisit/traffic.htm

National Park Service. (n.d.). *Annual park recreation visitation (1904—last calendar year): Arches.* NP. Retrieved from https://irma.nps.gov/STATS/SSRSReports/Park%20Specific%20Reports/Annual%20Park%20Recreation%20Visitation%20(1904%20-%20Last%20Calendar%20Year)?Park=ARCH

Parks Canada. (2002). *The Parks Canada mandate and charter.* Government of Canada. Retrieved from https://www.pc.gc.ca/en/agence-agency/mandat-mandate

Parks Canada. (2019). Introduction. *Government of Canada.* Retrieved from https://www.pc.gc.ca/en/pn-np/introduction

Parks Canada. (2021, March 3). *Jasper National Park: Avoiding crowds. Government of Canada.* Retrieved from https://www.pc.gc.ca/en/pn-np/ab/jasper/visit/foules-crowds

Redko, V. Y., Krasnikova, N. O., & Krupskyi, O. P. (2022). Overtourism effect management in destinations. In M. Valeri (Ed.), *Tourism risk: Crisis and recovery management* (pp. 199–219). Emerald Publishing Limited. https://doi.org/10.1108/978-1-80117-708-520221014

Rice, W. L., Reigner, N., Freeman, S., Newman, P., Mateer, T. J., Lawhon, B., & Taff, B. D. (2022). The impact of protective masks on outdoor recreation crowding norms during a pandemic. *Journal of Leisure Research, 53*(3), 340–356. https://doi.org/10.1080/002222 16.2021.1981791

Robinson, K. (2020, September 6). 'It's just craziness out there': Tofino and Ucluelet urging visitors to respect COVID-19 protocols. *Global News.* Retrieved from https://globalnews.ca/news/7319019/its-just-craziness-out-there-tofino-and-ucluelet-urging-visitors-to-respect-covid-19-protocols/

Romagosa, F., Eagles, P. F. J., & Lemieux, C. J. (2015). From the inside out to the outside in: Exploring the role of parks and protected areas as providers of human health and well-being. *Journal of Outdoor Recreation and Tourism, 10*, 70–77. https://doi.org/10.1016/j.jort.2015.06.009

Ross, W. (2021, September 14). Now is the time to go to normally crowded Banff. *The Daily Beat.* Retrieved from https://www.thedailybeast.com/banff-is-normally-crazy-crowded-now-is-the-time-to-go

Shelby, B., Vaske, J. J., & Heberlein, T. A. (1989). Comparative analysis of crowding in multiple locations: Results from fifteen years of research. *Leisure Sciences, 11*(4), 269–291. https://doi.org/10.1080/01490408909512227

Siegler, K. (2021, July 9). *An explosion in visitors is threatening the very things national parks try to protect. NPR.* Retrieved from https://www.npr.org/2021/07/09/1014208160/national-parks-crowds-litter

Stanfield, R., Manning, R., Budruk, M., & Floyd, M. (2005). *Racial discrimination in parks and outdoor recreation: An empirical study.* The 2005 Northeastern Recreation Research Symposium, 247–253. Retrieved from https://www.fs.usda.gov/ne/newtown_square/publications/technical_reports/pdfs/2006/341%20papers/stanfield341.pdf

Stilgoe, J. R. (2001). Gone barefoot recently? *American Journal of Preventive Medicine, 20*(3), 243–244. https://doi.org/10.1016/S0749-3797(00)00319-6

Suana, I. W., Ahyadi, H., Hadiprayitno, G., Amin, S., Kalih, L. A. T. T. W. S., & Sudaryanto, F. X. (2020). Environment carrying capacity and willingness to pay for bird-watching ecotourism in Kerandangan Natural Park, Lombok, Indonesia. *Biodiversitas Journal of Biological Diversity, 21*(5). https://doi.org/10.13057/biodiv/d210557

USA Today. (2021, June 12). *Record number of visitors flocked to Yellowstone National Park in May.* Retrieved from https://www.usatoday.com/story/travel/experience/america/national-parks/2021/06/12/yellowstone-national-park-record-number-visitors-may/7668321002/

Wagar, A. J. (1964). The carrying capacity of wild lands for recreation. *Forest Science, 7*, 1–24.

Wall, G. (2020). From carrying capacity to overtourism: A perspective article. *Tourism Review, 75*(1), 212–215. https://doi.org/10.1108/TR-08-2019-0356

Weber, M., Groulx, M., Lemieux, C. J., Scott, D., & Dawson, J. (2019). Balancing the dual mandate of conservation and visitor use at a Canadian world heritage site in an era of rapid climate change. *Journal of Sustainable Tourism, 27*(9), 1318–1337. https://doi.org/10.1080/09669582.2019.1620754

Winter, P. L., Selin, S., Cerveny, L., & Bricker, K. (2019). Outdoor recreation, nature-based tourism, and sustainability. *Sustainability, 12*(1), 81. https://doi.org/10.3390/su12010081

Wiyono, K. H., Muntasib, E. K. S. H., & Yulianda, F. (2018). *Carrying capacity of Peucang Island for ecotourism management in Ujung Kulon National Park.* IOP Conference

Series: Earth and Environmental Science, 149, 012018. https://doi.org/10.1088/1755-1315/149/1/012018

Xiao, Z., Henley, W., Boyle, C., Gao, Y., & Dillon, J. (2020). *The face mask and the embodiment of stigma. OSF Pre-Prints.* https://doi.org/10.31234/osf.io/fp7z8

Zelenka, J., & Kacetl, J. (2014). The concept of carrying capacity in tourism. *Amfiteatru Economic, 16*(36), 641–654. Retrieved from https://www.econstor.eu/handle/10419/168848

Zhang, J., Cheng, Y., & Zhao, B. (2021). How to accurately identify the underserved areas of peri-urban parks? An integrated accessibility indicator. *Ecological Indicators, 122*, 107263. https://doi.org/10.1016/j.ecolind.2020.107263

6 Comparison and Enlightenment of Leisure Sports Resorts in China and Abroad

Ping Ling and Yingqi Zhang

Introduction

In today's world, the tourism industry is undergoing a development process, moving from sightseeing tourism to vacation tourism to experiential tourism. Vacation tourism has become the most important format for the development of tourism. In the research report on the development of the global tourism industry, the World Tourism Organization (WTO) Secretariat highlighted that the changing trend of tourism products will follow the patterns of vacation tourism products, special tourism products, and personalized tourism products.

China's high-end holiday tourism products have developed rapidly in recent years, but it is still in its infancy. Research on the market demand for high-end holiday tourism products has not been extensive. Current research focuses on specific tourism projects, how to operate a resort as a business format, how to make the resort meet customer needs, and how to improve quality from management to product design. Nowadays, new products that use leisure sport projects as resort areas, such as "golf clubs", "yacht clubs", and "equestrian clubs", have been favored by the market. Therefore, how to improve service quality from business philosophy, management methods, and product design has opened new opportunities. This new type of resort development model has positive significance. Sport resorts will improve the quality of high-end leisure sports resorts and may become a trend in the development of resorts in China. For that reason, this chapter takes the Mediterranean Club and Jiulongshan Resort as the research foci, and through comparative research aims to provide a reference model for the development of leisure sports resorts in China.

The Concept, Classification, and Basic Format of the Resort

Holiday is interpreted in English-Chinese dictionary as "to spend one's holidays", that is, people use the activities and methods of holiday for the purpose of recreation and sports. The concept of vacation has existed since the Roman period. It was developed from public baths. Vacation-tourism is the main purpose and content of vacation and leisure, which is used to relax the mind and body.

In 500 BC, tourist resorts first appeared in Europe, and the world's first resorts appeared in hot springs and mineral springs in Greece. The development of

DOI: 10.4324/9781003476658-8

resorts as an industry in Europe dates back to the 16th century in Belgium. There are a lot of mineral springs in a small town called Spa. People in the town find that bathing or drinking spring water and mineral water will have a beneficial effect on the body. This attracted people around the town to visit Spa to treat disease. So, the town's name, Spa, has become synonymous with hot springs, mineral baths, or spas.

After the Second World War, the rapid development of science and technology has greatly improved productivity. People have improved and increased their income levels and leisure time, especially the emergence of paid vacations, which has created good conditions for people to spend long vacations. In this case, tourist resorts have experienced unprecedented development.

A resort refers to a group of buildings used for leisure and entertainment, usually operated by an independent company, but also operated by several groups in cooperation. In order to allow tourists to enjoy their perfect vacation, the resort usually has a number of entertainment facilities available to meet the various needs of guests, such as food, accommodation, travel, tourism, shopping, entertainment, recreation, sports, and health. Some resort towns are also officially titled as resort towns. Basically, the term "resort" is used to distinguish it from hotels that stand alone and do not provide resort facilities. However, hotels used as accommodation are also an important part of the resort.

The WTO defines a resort as:

–   destinations that provide entertainment facilities that attract guests,
–   provide food, accommodation, and other services for out-of-town guests,
–   provide various services that can make guests stay longer.

According to the classification of accommodation facilities, resorts can be divided into different categories, resort hotels being the most common form of resorts. Resort hotels are often based on natural resources (such as beaches, lakes, mountains, forests). By providing recreational facilities and services, it creates a pleasant, peaceful, and relaxing environment for tourists and meets the needs of tourists for leisure, sports, and entertainment.

The history of vacation tourism in China begins very early. Early resorts were built for high-ranking nobles. The most typical ones are royal gardens and private garden-style resorts, while modern resorts are mostly concentrated on seashores and mountains. Spas and health resorts are mainly for leisure and wellness.

After clarifying the concept, we can categorize the resorts. Depending on the content of the park, the resorts can be divided into Integrated Resorts and Town Resorts. According to the resources, the resorts can be divided into mountain areas, seaside resorts, golf resorts, gambling resorts, and cultural resorts (churches, museums). According to the main project, the resort can be divided into mountain outdoor sports resorts (mountain climbing, rock climbing, mountain biking, hiking), ski resorts (various snow activities), golf resorts, coastal resorts (yachts, swimming, sailing, surfing, paddling, diving, canoeing, fishing), wellness resorts (spas), and ecology-type resorts (RV camping).

Leisure life has become a developing trend. The increase of leisure time, improvement of leisure facilities, change of life concepts, and the autonomy of activities has laid a foundation for people to enjoy leisure better. During weekend leisure activities of various forms, rich in content and changing with each passing day, we seem to find the following basic characteristics:

- *Evolution from indoor to outdoor.* The past leisure methods of the Chinese nation originated from farming civilization. Among the farming civilization, people's cultural leisure was mainly dominated by piano, chess, calligraphy, painting, and leisure sports focused on yoga and qigong. Nowadays, people's daily lives are more concentrated in big cities, so there is a need to be able to reintegrate into nature, society, and beauty.
- *Evolution from static to dynamic.* The ancients' piano, chess, calligraphy, and painting leisure activities are static activities. In the modern pace of life, more investors are willing to set up resorts in areas such as coastal and mountainous areas, and organize multi-elements, such as mountain climbing, skiing, surfing, golf. These holiday activities have the dynamics to improve the physical fitness of tourists.
- *From day to night.* The quality of night activities has become an important factor for visiting residents.
- *The evolution from passive viewing to active participation.* As far as sports are concerned, there are many sports that are ornamental but more inclined to entertain themselves. For example, from a modern perspective, table tennis and badminton may not be able to welcome a large number of spectators in China. However, in various cities, different table tennis clubs are generally overcrowded, so these sports have brought healing health attribute. It can be concluded that today's ordinary people have the intention to focus on their surroundings and are willing to maintain their health through physical exercise, green roads, and slow trails around them.
- *Evolution from elite to public.* In the Chinese market, due to the lack of land resources, golf activities still belong to the elite. At present, the number of golf courses approved by the state in China is about 600, while the number of golf courses in the United States can reach more than 25,000. The golf club membership fee in China ranges from 600,000 to 800,000. In the United States, its annual fee can be as low as $400, so golf has become popular in developed countries, such as the United States.

The Mode of Market-Oriented Operation of the Mediterranean Club Resort

Club Med was founded in France in 1950 and currently has more than 80 resorts around the world. It is the only chain sports resort in the world. The main feature of these resorts are leisure sports. The development model of tourism products, such as the Mediterranean Club, is the earliest and most successful development model, and it has gradually become the trend of the development model of today's resorts. Its main feature is that it completely relies on market mechanisms and upholds an excellent business philosophy and management system. In addition, Club Med actively innovates to shape the characteristics of sports. With sports having their own characteristics, they attract tourists from all over the world. Club Med is a global resort group developed through chain management. There are currently five Mediterranean clubs in China, distributed in Harbin Yabuli, Hebei Beidaihe, Hainan Sanya, Guangxi Guilin, Zhejiang Anji, and other places. These locations are separated into different types, such as Guilin being a landscape type, Beidaihe and Sanya are coastal types, and Jabri is a type of snow mountain.

As a world-renowned group, the Mediterranean Club's first business strategy is to think that consumption is a loss for holidaymakers where every consumption decision, such as the choice of restaurants, the consumption of shopping malls, participation in skiing, surfing, and diving activities is made in the resort area. If these decisions occur, any consumption in the resort area is lost, and then the form of vacation is destroyed.

The second business strategy is "one-price all-inclusive", that is, if all consumption in the resort is one-price all-inclusive, then the status of each consumer in the resort is equal, so consumers do not need to make consumption choices. All areas are members. The services in the area include the provision of a wide variety of sports activities, different types of catering, and cost-effective vacation products, mainly leisure sports.

In addition, the Mediterranean Club also considers parent-child tourism and children's characteristics. It has a corresponding brand program for children from 4 to 11 years old, 11 to 17 years old, and adults. One of the main features is "soft service", which is called G.O. by the public. G.O. mainly refers to the employees in the resort area. Its employees are both waiters and restaurant consumers, coaches and co-trainers, as well as guides and playmates. Staff can teach tourists to snorkel or accompany tourists to enter the 12 m, 15 m.

At present, the well-developed resort areas in China include Jiulongshan Resort, which is mainly characterized by leisure sports. There is sailing, windsurfing, equestrian, horse racing, golf, hot air balloon rides, water parks, and other projects in the resort area, all highly interesting. For example, the outermost area of the resort arena is F2 racing, the middle lane is horse racing, and the inner is equestrian venue. At the same time, the positioning is accurate. The distance by driving car is one hour from Shanghai, one hour from Suzhou, one hour from Wuxi, one hour from Hangzhou, and one hour from Ningbo. The main consumption targets are the five major cities surrounding the resort. The equestrian consumer groups of Jiulongshan Resort in Jiaxing are officials of the Shanghai Consulate and foreign tourists, with a leading marketing concept. At the same time, it also has an excellent foreign management team, establishes a commercial platform, and fully integrates with major events. The current construction style is completely the style of a five-star hotel in Venice, Italy.

A comparison of the Mediterranean Club and the Jiulongshan Club reveals the differences and connections between them. The first is the membership system. Both the Mediterranean Club and Jiulongshan Club have golf courses. The membership fees of the Mediterranean Club are low, unlike other domestic golf clubs, which have expensive membership fees. The second is the supporting entertainment.

The Mediterranean Club is equipped with beach volleyball, beach football, water polo, tennis, diving, sailing, windsurfing, water skiing, and other activities. Jiulongshan Holiday Club features equestrian, yacht, and other activities, but the supporting facilities are relatively single. Currently, the Jockey Club has become a key project in the resort area, which will provide a variety of training with good development potential. Finally, it is a source of income. The Mediterranean Club has a large number of loyal customers for repeated vacation consumption, and the levels are concentrated. In the mid- to high-end segment, Jiulongshan Club's main profit channels are sales of membership cards, real estate operations, and commercial activities.

First of all, the current resort area has largely got rid of a single operation, and has continuously developed various leisure, entertainment, and fitness products. Therefore, investors need to focus on two points in the development of the resort area: the entrance of the product (what product can the resort provide), and the entrance of traffic (how many customers can the resort have and how many of them are repeated customers).

Secondly, the project needs to target people with a strong social foundation and a material foundation. At present, the marketing plan of China's resorts is mainly aimed at customers who have gradually formed mid- to high-end markets in China. However, there have not been any high-end sports resorts integrating golf, polo, yachting, aviation, racing, and motorcycle in China. The development direction and core competitiveness of resorts should be closer to the middle- and high-levels customers.

Third, from the perspective of sports design, there must be a certain connection between resort projects. The combination of "golf + yacht" has been used in many resorts. Very successful cases include vacations in Los Angeles, San Diego, Miami, etc. After the comparative study, it was found that there is a very close relationship between yachting, sailing, and golf. Many golf courses in the United States are built by the sea. The wealthy yachts and sailing boats also generally rely on the Gold Coast in the rich areas of San Diego. The projects are in line with the vacation demands of the rich, and projects such as polo, aviation, racing, and motorcycle are not available in other resorts.

Finally, the design of the Mediterranean Club's sports programs is more geographically focused. Features as a reference, highlighting "the theme of a certain sport, the characteristics of the integration of multiple sports events", market segmentation to meet the needs of each vacationer, so that each resort can propose an improved sports project design program.

The Basic Strategy for the Adjustment of the Resort Format in China

For China to successfully construct resort areas, seven issues must be considered in particular:

1   How do resort products carry out top-level design and product positioning?
2   What kind of product system and content are in line with new consumer demands?

3 How can management maintain resort software? What produces return customers?
4 How do intellectual property products become explosive derivatives?
5 How can the resort companies create business value? What about on weekdays? How can management solve the problem of low season?
6 How can resorts extend the stay and increase the unit price of tourists?
7 How does experience design work? How can you promote more secondary matches from the experience?

By thinking about these issues, we can provide some reference ideas for the development of resort areas in China. These ideas will cultivate customer loyalty with the help of surrounding attractions as supporting projects.

The Mogen Mountain in Deqing County, Huzhou City, and Zhejiang Province has many tourists attraction. The earliest group attracted to the Mogen Mountain area was foreigners traveling in Shanghai. In fact, the main reason for attracting this group was the unique atmosphere of the mountains and water around Mogen Mountain.

Hot springs and spas are important supporting facilities in the resort areas, attracting customers during all four seasons. More famous are the hot spring resorts in Hokkaido, Japan. The earliest known spas started in Greece, later leading to the accelerated development of spas development in Spain, Italy, and other countries. In China, Kouquan Hot Spring is also accelerating its development and trying to form a package with the resort.

The format and proportion of large-scale resorts is a routine solution to the problem of the off-peak season. For example, Tianhuangping Scenic Area, Huzhou City. Within a range of 50 km distance in Hangzhou, it can maintain three months of ice and snow due to high altitude. Therefore, there is a three-month ski season in winter. It can offer a series of mountain outdoor sports, so it can satisfy the need for outdoor activities in different seasons, such as skiing and skating in winter, and surfing and grass skiing in summer. These activities are a much better way to solve the seasonal differences.

The operation mode of the camp and the nature education curriculum system are effective measures to attract passengers flow of the resort and solve the off-season problems. In particular, youth training and Communist Party of China special education are examples of camp education. Additionally, team dynamic training in China and RV camping activities in developed European countries and the United States can also be considered as camp education. Camping resorts can be used as a reference and some off-season collective projects can be launched.

Design an Immersive Experience

Casinos in Macau and Las Vegas are considered an immersive experience. Today's active and respectable immersive experiences include ice and snow activities and golf activities. There are no two identical golf courses in the world. Thus, visitors can experience the characteristics of 100 golf courses by participating in 100 golf events. Most golfers choose golf experience because they love golf travel.

Brand design and product development need to analyze the resort design to understand which scenes are narrative and which are lyrical. For example, there are many resorts that can set up special Love Story telling to integrate elements related to love and create a cultural atmosphere through a bonfire party. The Happy Valley in Shanghai Laoshan Resort can create water display and design them according to the song lyrics.

Marketing strategy, in the context of the macroeconomic downturn, begins with physical and mental health and quality of life. The 2019 tourism index of major cities in China, excluding first-tier cities such as Beijing and Shanghai, has experienced varying degrees of decline. Under the premise of such a decline, and with human-centered goals, the development prospects of tourism activities must be closely related to the improvement of physical and mental health. Therefore, resort areas need to start with physical and mental health and focus on quality of life to carry out related marketing and promotion, which are important, realistic pursuits.

7 Winter Sport Tourism

An Ecological and Economic Point of View

Rudolf Leber

Introduction

The commercial use of winter sports has long since found its way into the touristic centers of winter sports regions. Despite this fact, commercial winter sports only target a certain audience, those which have the opportunity and interest to engage in sport activities in the alpine terrain and open nature (Figure 7.1). A considerable part of the population has no access to these activities due to the financial expenditure of skiing and winter sports as well as the lack of desire to engage in sport and exercise in nature (Arbesser et al., 2008; Steiger et al., 2019).

It is essential for municipalities (communities, cities, etc.) in the field of winter sports tourism to provide optimal conditions in all their facets for the guest. As the responsible institution, municipalities have a very significant function to fulfil here. Developments (in winter tourism) must be effectively managed while remaining shaped by the existing framework conditions of the economic environment, and mediation between economic forces and forces of socialization and integration must occur in a determined way.

The central theme (winter tourism) is to predict the development of winter sports for the next ten years with socialization, integration, and cultural work playing an essential role in the development of our adolescent youth. Based on a long-standing tradition, the municipality (communities, cities etc.) has always been committed to create establishments and institutions, especially for the general population, due to its self-perception and the ability to improve their education and quality of life. In the future, it will, therefore, be a priority for all involved in this field to secure their important social and integrative status to address the tension of municipal challenges. In this way, concrete movement-oriented behavior patterns including exercise in nature, social learning, and diverse experiences can become internationalized in the field of winter sports.

By training the ability to concentrate, researchers can detect an increase in the cognitive development process, demonstrating the importance of winter sports. If children and young people are to properly learn the key competences of "independence, creativity, problem solving", they need appropriate educational approaches, such as exercise, creativity, music, and social learning. These approaches will enable optimal support, which is becoming increasingly important in

DOI: 10.4324/9781003476658-9

Figure 7.1 Alpine Terrain in Winter.
Source: Rudolf Leber.

our highly complex society. In concrete terms, the 'Campus educational establish-ment' project (school pilot project) is concerned, inter alia, with teaching-specific life processes which are intended to serve as a starting point for children to gain as much experiences as possible and to develop their unique personalities. Due to the psychological and physical resilience of the children (personality development), this comprehensive approach increases their ability to learn how to use their free time in a constructive way.

Target Groups of Winter Sports

Regarding the topic of winter sports in schools, it can be said that school sports are currently in a quandary (Müllner, 2013). Physical education is usually further rel-egated to the background of school autonomy (e.g., cancellation of project weeks, school ski courses). Yet, for children and young people in particular, exercise, i.e., sport, is of utmost importance. Through sport, educators promote not only an aspect that is beneficial to general health but also an important factor in the socialization process. During sport, mental and physical well-being may go hand in hand, leading to an increase in the quality of life. Student's physical abilities are sources of con-structive self-affirmation and, thus, students can positively develop their personality.

Various studies in recent years demonstrate the importance of daily exercise and sport, suggesting that physical education should no longer be a secondary mat-ter in schools. The cutbacks in school sport budgets must be regarded as a stark contradiction to scientific findings and, thus, there is a great need for action by the Ministry of Education. Physical activity is one of the fundamental ways for children and young people to express themselves. Simply for the fun of the activities and the experiences they create, children and young people use every opportunity to engage in physical activity. Through physical activity and play, they learn to manage problems and explore a wide variety of operating methods. Hence, physical activity offers diverse experiences and the opportunity for a holistic education and upbringing (cf. Leber, 2014).

During winter sport weeks, the opportunity at hand is to use school pedagogues and ski instructors in an interlinked manner. While this enables both groups of teaching staff to gain insight and experience in the other's field, it also provides students with the opportunity to choose from a very wide range of programs based on their desired free time. In order to achieve this, experienced educators are needed. Modern leisure education offers the best conditions for the development of autonomy and self-responsibility. It promotes tolerance, mutual respect, and a considerate approach to one another while supporting the integration of children with a migrant background. During winter sport activities, children have the opportunity to gather varied experiences through self-determined leisure activities, to reflect on these, and subsequently to live an increasingly self-determined life. In this way, children learn to differentiate between meaningful, satisfying offers and consumption-oriented activities as well as experience active engagement and relaxing idleness as pleasurable and enriching. Since the greatest attention is paid to those leisure activities which – actually or allegedly – are new, the employment of experienced snow sports educators provides an important influence in the care of children and young people, especially in the areas of sports, music, and creativity.

Various studies have found that children and adolescents enjoy devoting their leisure time to winter sports and that the experience of nature, rest, and relaxation are also particularly important. As part of their winter sport activities, children want to be affected by as few disruptive factors as possible. Thus, there should be a limit to the number of people allowed on the ski slope, the waiting times at the lifts should be relatively short, and the quality of the ski slope should be high. Moreover, children and young people prefer to ski with friends. Skiing must be designed accordingly for children and youth to be excited for winter sports. Furthermore, it has been observed that, for children and teens, the entry into winter sports occurs mainly through family and friends. Age of entry for children is about three to five years old and for teenagers (with the beginning of school winter sports weeks). Therefore, the interest of families and children must be prioritized to ensure they continue to participate in winter sports for their own well-being.

In addition, many people practicing winter sports consider participating in school ski courses (or winter sports weeks) as the starting point for a future of winter sports engagement (Figure 7.2). Implementing winter sports weeks as part of the general school education is a long-standing tradition in Austria. Due to economic problems (e.g., single earning families), summer sports weeks had higher participation rates in comparison to winter sports weeks. Thus far, efforts to raise interest for winter sports have failed, owing to the economic demands placed on parents and schools.

Figure 7.2 Winter Tourism in the Alps.
Source: Rudolf Leber.

The Development of Winter Sports

The development of winter sports tourism in the European Alpine regions of Austria has been very successful in recent decades. The climatic prerequisites, scenic topographical position, European hospitality, and enthusiasm for winter sports result in excellent conditions. The success of this destination reflects the economic dynamism in this sector (winter sports) which has been a constant feature for the past half a century. However, appearances can be deceptive. Alpine associations have been raising a concerned voice for years, drawing attention to the consequences of climate change and the resulting snow-poor winters. These associations have also warned of problems related to landscape protection, as well as the demographic ageing process in Europe, causing a reduction in the skiing population.

Particularly in the field of winter sports, Austria is one of Europe's most attractive destinations in the world. Due to the significant price increases in winter tourism, winter vacations and school skiing courses are also declining. Nevertheless, about 15% of the Austrian population still takes a one-week skiing holiday and ski touring is currently experiencing a boom again. The importance of proper ecological use in the Alps has become evident from the increasing number of climate-related storms in recent years. Therefore, it is essential that the tourism and winter sports industry operate in a mild and responsible way.

There is certainly no doubt that significant strain is placed on the environment, especially when the consequences of civilization and technical progress are noticeable in nature. The natural balance of Earth has been, and continues to be, disrupted by overpopulation and claims for high standards of living, causing serious problems. The limits are visible. This is particularly evident in an ecosystem as important and delicate as the Alps (Vigl et al., 2016). The ecological developments, agricultural, forestry use, and tourism in the Alps are risk factors that endanger natural functioning.

The forest's illness makes this particularly clear and shows how precarious the present state of affairs has become (Weiss, 2001). The loss of forests would not only destroy the basis of many Alpine countries, it would also affect, via hydrology, the areas of all river systems that have their origin in the Alps. The consequences of forest loss include constant avalanches, karstification (Kiraly, 2003), unchecked water discharge, and floods. This demonstrates that it is in the interest of the general public livelihood to keep our environment as intact as possible.

As explained beforehand there are many hazard areas, it is important to know their scale in terms of quantity and quality, as well as their interdependencies. The overall danger cannot be dealt with by looking at the other regions in an accusatory manner. Instead, it is essential first and foremost to do everything possible in each respective field to reduce the specific burdens. The worried voices of Alpine organizations in Austria are pointing to new paths in winter tourism, as winter sports in the Alps are adapting poorly to climate change. There is a danger of less snow, fewer guests, and an economic catastrophe. In fact, over the last three decades the number of overnight stays in the winter season has doubled to around 60 million, while the occupancy rate of available beds has risen to more than 30% in the same period. The number of overnight stays by foreign guests continues to rise considerably. In the last years alone, the number of overnight stays by foreign guests has reached around 50 million from November to April. However, locals are, again, increasingly appreciative of the quality our local winter tourism offers. Despite all criticism, Austrian winter sports tourism has remained thriving to date. The organizational task of the future is to develop a stronger market for Austrian winter sports tourism in a constructive and creative way.

Success in winter sport tourism is the result of joint efforts from the entire nation, tourism associations, mountain railways, hotel industry, gastronomy industry, trade industry, and local populations. The affinity with the population engaged in winter sports, the interest groups, as well as the professional image cultivation also contribute to the success of winter sport tourism in Austria. The overall economic importance of winter sports is usually underestimated. This is primarily due to the fact that winter sports are not a separate statistical economic sector in its own right, but rather it is made up of a wide range of industries and economic sectors. It is undisputed that winter sports are of particular importance in Austria and that they produce considerable contributions to Austria's economic performance. While winter sports are generally associated first and foremost with winter sports tourism, ski schools and perhaps also the ski industry, sporting goods retail, media sectors, mountain railways, and sports associations (above all alpine organizations) are hardly taken into account.

The fact that the importance of winter sports extends far beyond winter sports tourism is reflected in the value creation network of winter sports, exemplified by the cableway industry and ski schools.

Anyone who has ever hiked through alpine ski regions on a beautiful autumn day will have witnessed the extent of the damage that winter tourism has already caused to nature in recent years and decades. Skiing in its present large-scale form burdens and changes the landscape, endangers the mountain forest, destroys vegetation, and pollutes the groundwater. All experts agree that a further exorbitant increase in winter tourism can be expected in the coming years.

Uses of Agriculture and Its Significance in the Alps

The present-day appearance of the Alps was largely shaped by agriculture. Until well into this century, it served the self-sufficient mountain population. When hardly enough land was available in the valleys, new land had to be sought constantly. Alpine pastures were created above the steep hillsides, where the cattle were taken up during the summer. This was only possible at the expense of the forest. Its protective belt over the valleys was constricted and partially destroyed. Based on previous reports, we know that this approach was radical and failed to understand the laws of nature as well as take the future into account.

It is clear that the natural forest line has been lowered by 200 m in large areas due to the plight of farmers and partly because of deforestation benefiting other sectors (e.g., mining). This assisted the farmers, but also increased risk by reducing the protection provided by the forest and created large areas of land with faster water runoff during heavy rainfall.

Structural change in the Alps began with the development of transportation, the growing number of new employment opportunities, and market competition for agricultural products. The difficult conditions of production are making the mountain landscape increasingly unattractive. At around 1,500, the number of mountain pastures has remained almost unchanged over the past 30 years. On almost 50,000 hectares, the same number of livestock is still being raised, although there have been changes and cows have frequently been replaced by young cattle due to the lower amount of maintenance they require.

Site-specific and careful management of both the valley areas and alpine pastures contribute to the preservation of the cultural landscape. For this reason, as many alpine pastures as possible are being preserved. If their utilization were to cease without afforestation, this would rapidly increase the risk of erosion and avalanches. Long grass that grows in place is not only an ideal sliding track for avalanches, but it can also be torn out in frozen bundles when glide avalanches push downhill, uncovering the ground and creating erosion gullies.

The existing grazing rights for livestock in the mountain forests and in the mountain pastures continue to pose a serious problem. The cattle are permitted to be driven into (they are allowed to graze there) the forest areas, thereby eating not only grasses and herbs but also other forest plants. Livestock also cause damage to the soil and the roots of the trees by trampling. The damage caused by trampling leads to compaction of soils, creating damaged roots. The damaged roots are a source for infection and tree diseases and the young tree plants that were eaten by the cattle are missing and can no longer rejuvenate the forest. The amendment of grazing rights is difficult, although the government of Austria is actively working to achieve this change.

The Importance of Forestry

In the high mountain regions, forests fulfil crucial protective functions. Without them, settlement in valleys would not be possible. The most important function is soil protection.

With their deep and far-reaching root systems, the trees keep the soil firmly in place, open it up (they supply the soil with oxygen and at the same time ensure continuous aeration) and ensure a constant formation of humus from foliage and coniferous litter. The canopy alone ensures that the precipitation does not fall to ground with full force, but is slowed down, dispersed, and immediately evaporates again.

The forest ground absorbs water like a sponge due to its structure and releases it more slowly than, for instance, meadows. It serves as a water reservoir and altogether prevents soil erosion as well as flooding caused by restricted water runoff, which frequently occurs in the sparsely wooded high mountain regions. In winter, the forest catches the snow in the canopies and tree trunks. It slows down defrosting and is the best protection against avalanches. Ultimately, the forest has a higher level of humidity as a result of constant evaporation and, through its shadow effect, has more balanced temperatures than the surrounding open field, providing good climate protection. Its protection against wind influence is equally important. In total, about 50% of the mountain forest performs a protective function!

It is worrying that the forests in the Alpine region are seriously threatened by forest disease, referred to as forest dieback. Science fears that the impact of emission and pest infestation could lead to the loss of vitally important forest areas. About 40–50% of the forest areas are affected in a visible way. In order to fulfil its function, the mountain forest requires a special composition based on tree species and a tiered structure. Silver fir, beech, maple, pine, larch, and spruce are the most important types of wood. In this respect it is essential that the presence of spruce is not excessive, as this is the most fragile tree species. It is shallow-rooted and particularly vulnerable to storms and snow.

In the past, various factors led to this tree species becoming increasingly dominant. Today it occupies approximately 70% of the surface area (in Austria mountainous regions), compared to approximately 50% in the past. The silver fir and the beech have declined dramatically. Undoubtedly, the high economic performance of spruce was decisive for this, as it can be regenerated most easily. Nowadays, it is clear that only a multi-layered, mixed forest can cope with the climatic and site conditions of the high mountains. For this reason, all forest management is being oriented toward this ecological goal.

A further problem is the difficulty of managing forests on steep slopes. For this purpose, forest roads and transport facilities had and still must be built.

However, these also bring about an ecological disadvantage by cutting into the slopes and changing the hydrology, leading to sealing, thus accelerating water run-off. With a path density of 20 running meters per hectare, this accounts for about 2% of the forest area.

It must be noted that the Alpine region forest will certainly not be able to cope with further damage. Moreover, this also sets clear limits to the development of skiing.

Hunting

The high mountains are home to animal species whose existence is threatened by humans. These include, above all, the mountain hare, capercaillie, black grouse, ptarmigan, and other bird species. These animals reside in the forest or in special biotopes.

However, alongside these species, there are also wild species whose numbers have increased through human activity to such an extent that they endanger vegetation, especially forests. These include red deer, roe deer, and chamois. While all these wild animals are particularly fond of feeding on young trees, eating their buds, and thus preventing them from growing, red deer also remove the bark of older trees, which can result in their weakening and premature dying. Another dangerous aspect of their food sources is that these wild animals prefer tree species of ecological importance such as fir, beech, and spruce, resulting in an unbalanced structure of the forest. Additionally, this type of wildlife no longer has natural enemies (bear, lynx, wolf, etc.), thus, they frequently have overly large populations that cannot be sufficiently reduced through hunting. Excessive wildlife populations have thus become a major damaging factor for mountain forests.

A close connection exists between the behavior of the wild animals and the constantly growing number of visitors to the mountain forests. The wildlife, which avoids contact with human beings, retreats to quieter areas and the elevated number of animals in these habitats lead to more wildlife-caused damage. The animals have also changed their living habits by becoming more active at night and, therefore, more difficult to hunt. However, this is not a sufficient excuse for the high numbers of wildlife, as this was already criticized before the pressure to regenerate existed. For the animals, the effects of human activity in their natural habitats can be particularly disruptive if their necessary quiet time is constantly disturbed and they are frequently driven away as a result. These impacts are higher in winter than in summer. The capercaillie, for example, has such a high energy consumption while fleeing that it can hardly cover its needs considering the lack of food in winter. Skiers must especially avoid causing these disturbances by following the designated and sign-posted ski runs and tracks. In the meantime, wild animals have gotten used to this.

Summer Tourism – Winter Tourism

Tourism is the youngest, but by far the most important, way to use the Alps. In Switzerland, for instance, tourist revenue accounted for 10% of all export earnings in 1985. The share of this industry, in terms of income earned, amounts to more than 90% in typical tourist communities. It has not always been this way. Until well into the past century, the remote location, weak agricultural sector, difficult working conditions, and poor infrastructure were responsible for poverty in large parts of the mountain population.

Tourism helped to overcome this hardship, which eventually expanded to an unexpected extent. The continuous, increasing demands for newer and better ways of spending leisure time has shaped this region. Originally, the natural environment was characterized by agriculture and later it became a recreational area. The benefits, however, also had negative impacts on the area, which, at first, were not realized.

For a long time, summer tourism used to be the focus in comparison to winter tourism (Figure 7.3). It is often overlooked that major infrastructure and tourist facilities were built mainly for the needs of summer guests. Roads, hotels, secondary housing, recreational infrastructure, hiking trails as well as waste disposal facilities were initially constructed for the summer season. The desire to make better use of the facilities and to employ workers on a more regular basis was frequently the driving force behind the development of a second tourist season in winter. This process was predominant in most Alpine countries, with the exception of France, where the ski resorts referred to as 'Retorted ski resorts' are now seeking opportunities for summer. Without the winter season and, thus, skiing, much of what is now perceived as a burden for the Alps would nevertheless exist.

Summer tourism additionally caused specific disruptive effects on the mountain regions. These areas are more accessible by foot and by vehicle in summer than they are in winter when forestry and agricultural tracks, hiking trails, and access roads lead almost everywhere. This access largely interferes with the tranquility in even the furthest corners of the region. A harmful aspect is the fact that many tourists do not follow the paths, causing damage when stepping on off-trail vegetation.

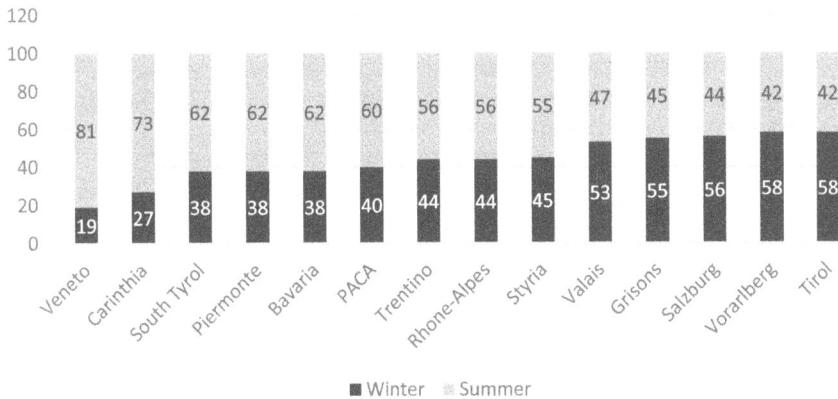

Figure 7.3 Seasonal Arrangement in Some Alpine Regions.

Source: Vanat (2018).

The consequences of erosion are visible in areas of high recovery rates. Not only rare animals are suffering from the disquiet, but rare plants are also affected when they are often taken home as souvenirs. New sports activities like mountain biking pose new problems.

The conflicting goals arise mainly from the large number of visitors in the Alps, 60% of which occur due to summer tourism. The corresponding traffic impact is, therefore, also to be attributed to this. Hardly any other environmental problem is currently the subject of such one-sided and generalized reports as skiing is, especially regarding the slopes. There was previously no reason to make a fuss about these problems, as many people were already visiting their preferred downhill slopes prior to the Second World War, but only a limited number of cable cars and lifts provided the desired ascent aids. Their construction did not begin to take off until 1950, rose steeply between 1960 and 1980, and has hardly developed further in Austria since then (Figure 7.4). Nearly equally important for the boom in skiing on slopes was the development of ski-run maintenance tools, which began to take off after the 1960 Squaw Valley Olympic Games. The smooth surface of the slopes made skiing attractive to beginners and advanced skiers.

Modern skiing requires appropriate slope facilities. Frequently, the natural shape of the terrain is used for the construction of ski runs; therefore, large area peneplanation and other changes to the ground have become commonplace. The existing plant cover and soil are pushed away, and a revegetation of the area causes great challenges and very high costs, especially above the tree line and over chalky

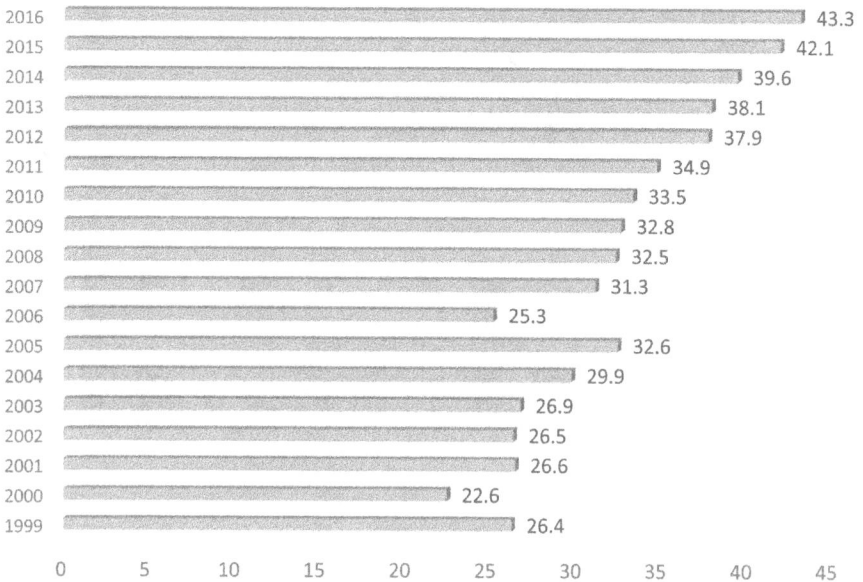

Figure 7.4 Changes of Cable Car Revenues in Austria from 1999 to 2006.
Source: WKO (2017).

subsoil. The compaction of the soil caused by the peneplanation, and the wide-open vegetation cover facilitates faster rainwater runoff and erosion processes, thus, it increases the risk of flooding and mudflows far down the valley.

The existing slopes require intensive management during winter. For this purpose, heavy equipment is applied, and a variety of chemicals are involved, especially prior to ski races. These interventions lead to heavy snow compaction and chemical pollution, resulting in prolonged snowy conditions in the spring and consequently negative impacts on surrounding areas.

In addition to energy, snow cannons also require large quantities of water, taken from the streams when they carry particularly little water. The use of snow cannons can also increase the amount of time slopes are used. For too long, the ecological condition of the slopes was not sufficiently considered. At first, nobody paid any attention to ecological stability. If there had been ecological guidelines right from the start, today's discourse would not exist. Thus, the above-mentioned shortcomings are blamed on skiing, as is the operation of the ski lifts, which is carried out without consideration of the natural environment. Particular attention must be paid to the opening of slopes when the snow conditions are not sufficient and in case of the irresponsible use of snow groomers.

However, it appears rather odd when critics, who rightly and loudly demand sufficient snow coverage, are the ones who are most strongly opposed to snowmaking systems. They argue that there is a need for fundamental reflection, and they fear that the rising growth rate of winter tourism will continue. Skiing mobilizes millions of people. Although there is a growing population and the problems increase, it is clear that each individual can contribute by skiing in a careful and responsible manner. Given the large number of skiers, the inappropriate behavior of only a small number of skiers creates a stress potential of considerable qualitative importance. Despite this, people still argue that the problems do not stem from skiers, wondering what else could they do except go up and down?

In addition to other forms of skiing, unrestrained and off-piste skiing poses particular problems. Many skiers find that the amount of activity on the ski runs has become too great and the slopes are too maintained and flattened. Frequently, skiers find this rather boring due to their improving maneuverability, leading them to move away from the 'highways' into the open country.

Relative to off-piste skiing, greater restrictions still need to be placed in the forest sector. The majority of wildlife lives in forests, due to the fact that they can find shelter for visual protection and a greater amount of food. The stress is all the more intense when off-piste skiers unexpectedly frighten the wild animals, and they are forced to flee. Furthermore, off-road skiers damage forest plants with their steel ski edges. These skiers are often unaware of it themselves because the young trees and bushes are hidden under the snow. Particularly in older forests, where the wide areas of trees attract off-piste skiers, young trees naturally grow and usually undergo a long phase of fortification. They are particularly sensitive to any kind of damage, and harm could occur if, in addition to the natural dangers such as animals' feeding habits, further negative impacts on the environment are caused by skiers.

Simultaneously, with the increased construction of ski slopes and an improving average level of participant skiing ability, good skiers have been able to switch to areas that were previously reserved exclusively for touring skiers. Therefore, the ski areas were used more extensively, as long as the terrain was suitable. In this way, however, the traditional forms of use that predominated (forestry, alpine, and farming) were further intensified and used for the purposes of sport.

With the rise of snowboarding, winter sports have been increasingly criticized by nature conservation associations than previously with traditional forms of skiing. This is due to the fact that new sports participants are now targeted when they would otherwise have been involved in sports activities in nature. The advertising in connection with snowboarding is a cause for criticism. In general, it is aggressively experience-oriented and lacks the information that describes how skiers can practice the sport in a sensitive natural environment.

It is not expected that sporting enthusiasts will demonstrate understanding for animals and plants when the general level of knowledge about ecological interrelationships is extremely low. With the disturbance of the wildlife in the forest, for example, by a group of snowboarders, the escape distance to possible areas where the wild animals could seek shelter has decreased, and that as a consequence, due to the elevated energy consumption of the wildlife, plants as food resources increase, and with it the environmental damages in the forest. The impact of wild animals eating plants as food is connected to a forest that is already struggling to survive. Therefore, the impact cannot be blamed on snowboarders, but nevertheless there is significant damage to the forest with lasting effect.

The problem of snowboarders' safety is less crucial than ecological climate, yet issues related to direct injuries or damage caused by frequent practice are worth mentioning. Snowboarders frequently have no alpine experience and yet they often snowboard in unsecured ski areas. Moreover, it must be considered that they cannot leave a slope by crossing it or ascending it slightly, as they would have to fight their way through deep snow on foot. This results in a conscious acceptance of the risk involved in using unsafe slope areas, as it could be too tedious to leave a danger area sideways.

Ski touring and mountaineering, regarded by many to be the gentlest form of skiing, becomes problematic if the ski mountaineer lacks basic knowledge of how to behave in nature. Considering the possible damages caused by skiing on slopes and off-piste skiing, the potential harm may seem insignificant. This is, above all, due to the fact that mountaineers develop a much more intensive relationship with nature than slope users. Those who ascend with their own strength and slowly discover the particularities of the landscape while ascending inevitably have a better understanding of natural habitats and can adapt to the laws of nature more easily than skiers using the slopes. Damage can also occur during ski mountaineering. Possible damage to vegetation and wildlife include:

- injury to the needles and branches of young trees,
- damage to shrubberies,
- damage to the sward,

- increase in browsing damage to vegetation, and
- disturbances of wild animals.

Since the total number of ski tourers is currently lower than that of slope users (the tendency is increasing – 30% in the last five years!) and the extent of possible damages is limited by these alone, elementary rules must be strictly followed:

- carpooling upon arrival,
- avoiding damage to grazing fences,
- avoiding areas of reforestation,
- protecting the ground vegetation in low snow conditions,
- avoiding wild animals and their feeding areas, and
- putting dogs on a leash in the forest.

The touring skier or ski mountaineer should be able to recognize the problems mentioned. This requires intensive education of all professional associations and schools, as well as guidance of the tourism industry, which I will certainly discuss in more detail in the following!

Evaluation

The utilization rates in the Alps shows that all of them have a considerable impact on nature. Even the careful use of agriculture and forestry is accompanied by risks for nature and landscape. Furthermore, there is a complex interdependence between all types of usage, some of which have positive outcomes. For instance, the benefits of tourism for farmers and the mountain population.

Although agriculture has lost income in some fields, this is offset by revenues from tenure (income from lease), room rental, and the possibility of additional income. These opportunities more than compensate for the disadvantages (due to reduced use of the slopes). Not to be overlooked are the negative consequences of tourism on forests, landscape, and hunting.

In all areas (in the optimal use of the alps) there is an abundant need and possibility to reduce and limit the specific burdens. This is the essential task, which can only be achieved if there is also coordination among the divisions. Any individual solution must also consider its impact on the overall picture to ensure it does not induce problems elsewhere. Networking applies not only in ecology but also in socio-economy. Cooperation is required for finding complete solutions (networking of scientific disciplines).

Possibility of Schools and Associations

Schools and associations (tourist and sports associations) have a crucial role to play in the emerging debate on the values of the recreational society. Schools and associations should set themselves more ambitious goals rather than satisfying individuals' short-term wishes. The sports clubs or schools where outdoor sports are practiced, in the interest of long-term preservation, must put the idea of nature protection at the forefront.

In clubs of a certain size and different age groups (from 5,000 members), knowledge about the problems of using of nature for sports accumulates and must be considered when making decisions. Only those who pursue long-term, idealistic values and ambitions can detach themselves from the commercially oriented superficial interests and make value-free judgments. With schools and associations, the possibility of informing the individual student or members about the available possibilities in combating climate change happens at community events, lectures, and excursions. The information shared motivates changes in behavior and increases interest in the preservation of nature or soft tourism.

To fulfill this educational mandate, the schools and associations that practice sport must develop (they will produce teachers from their own ranks) professionally trained teachers and produce informative material. The importance of skiing and nature preservation as a research subject should be treated as a priority. In addition, it is important that research and media does not give the impression that Austrians are overreacting to the demands for environmental protection. For example, the slogan "Skisport – Alpenmord" (Skiing – Alpinemurder) or "Ski Heil – Berg kaputt" (Ski Heil – Mountain destroyed) has been coined to point out this problem.

More serious than the buzzwords quoted is that schools refrain from offering skiing courses in the future to avoid disturbing the ecological balance of the Alps. Even though such schools are still the minority, the tendency is unmistakable. It must be assumed that in view of the unsuccessful nature preservation efforts in essential areas, a sham success is being pursued in tourism in general and in sport.

The opponents of sport in the open countryside derive from the incorrect assumption that the practice of sport in the countryside irrevocably consumes the countryside. To contrast this, the athletes, and their associations (schools) should emphatically make nature protection their own concern and make demands for nature preservation! The aim must be to win over the athletes in a positive way to encourage a commitment to nature conservation. This could also be a chance to draw attention to the big problems of environmental protection. One way of awakening the willingness of our fellow citizens to do this is to sensitize them to small, manageable problems such as skiing. Only those who know nature and have learned to love it will also work for its preservation.

Let us, therefore, try to contribute by transforming the people entrusted to us in clubs, schools, and ski lessons into active sport participants and nature protectors, who, based on their understanding of nature, actively support an environment worth living in and do not forget the human being!

Tourism Investments in Snow Sports

The value added by winter sports amounts to approximately €8 billion annually. Including the indirect and induced effects, the annual value-added contribution even exceeds €11 billion. Thus, in the total Austrian gross domestic product, the share of winter sports is about 5% and corresponds in an order of magnitude to the value added by retail trade, public administration, or business-related services (SpEA, 2013). Winter sports tourism accounts for the largest share, in the hotel, catering, transportation, sport, entertainment, and retail industries, generating around €6 billion in direct value added annually (SpEA, 2013).

Strategic Future Planning in the Winter Sports Sector

The development potential of winter sports in the core Alpine countries (Italy, France, Switzerland, and Austria) is in danger of geological erosion and needs to be revised with regard to its content (spirit – "A sports facility that, by linking science, research, teaching and sports medicine, rekindles and spreads the fire for sporting movement and performance in a municipality and beyond national borders, will help top athletes achieve their top performances nationally and internationally support and accompany the competition. This gives the youngsters a perspective and leads popular sport into a better future". cf. Leber, 2017) (Unbehaun et al., 2008). The following points should be considered when developing future strategies:

- Defining leisure-policy objectives (is in the realm of science);
- Resource management within the own sphere of action (tourism industry, municipalities, etc.);
- Emphasizing quality before quantity for the campaign image;
- Location and infrastructure management (planning etc.);
- Create programs of measure to make winter sports affordable again (e.g., subsidizing lift costs and equipment costs);
- Using the interest of winter sports tourism to promote the individualization of offers;
- Aim to achieve an increase in added value per unit area in the Austrian winter sports destinations, despite the background of climate change;
- Focus on high-quality, trained educational staff and sports teachers for summer and winter (in schools and clubs);
- Develop specific winter sport offers for all age groups to enable the development of new market opportunities;
- Implement appropriate support measures for school ski courses, for the pursuit of winter sports activities;
- Create opportunities to engage in winter sports activities at the primary school level in cooperation with ski schools;
- Adding an extended snow sport instructor training program should be considered at the university level;
- Supporting and implementing innovative ideas and projects.

The aim of economically and ecologically oriented leisure and sports management is to incorporate resources and preferences in a comprehensive way for the development of the leisure sector and thereby creating optimal conditions for winter and summer tourism.

References

Arbesser, M., Borrmann, J., Felderer, B., Grohall, G., Helmenstein, C., Kleissner, A., & Moser, B. (2008). Die ökonomische Bedeutung des Wintersports in Österreich. *Studie im Auftrag der Initiative "Netzwerk Winter ".*Wien: IHS Institut für Höhere Studien.

Kiraly, L. (2003). Karstification and groundwater flow. *Speleogenesis and Evolution of Karst Aquifers, 1*(3), 155–192.

Leber, R. (2014). *Planungs- und Facility management – im Bereich der kommunalen Bildungs- und Sportentwicklung.* Sportwissenschaft – Bildungs-, Schul- und Sportzentren. Band 2, Holzhausen – Verlag, Wien. ISBN 978-3-902868-81-7

Leber, R. (2017). *Aspern Sports Area – ein visionäres Spitzensportzentrum. Planung eines Forschungs-, Trainings- und Wissenschaftszentrums im Bereich der kommunalen Sportentwicklung.* Band 3, Holzhausen – Verlag, Wien. ISBN 978-3-902976-56-7

Müllner, R. (2013). The importance of skiing in Austria. *The International Journal of the History of Sport, 30*(6), 659–673.

SpEA – SportsEconAustria. (2013). *Ökonomische Bedeutung des Wintersports in Österreich.* Institut für Sportökonomie, Wien. IHS. Institut für Höhere Studien, Wien.

Steiger, R., Scott, D., Abegg, B., Pons, M., & Aall, C. (2019). A critical review of climate change risk for ski tourism. *Current Issues in Tourism, 22*(11), 1343–1379.

Unbehaun, W., Pröbstl, U., & Haider, W. (2008). Trends in winter sport tourism: Challenges for the future. *Tourism Review, 63*(1), 36–47. https://doi.org/10.1108/16605370810861035.

Vigl, L. E., Schirpke, U., Tasser, E., & Tappeiner, U. (2016). Linking long-term landscape dynamics to the multiple interactions among ecosystem services in the European Alps. *Landscape Ecology, 31*(9), 1903–1918.

Vanat, L. (2018). *International report on snow & mountain tourism – Overview of the key industry figures for ski resorts* (10th Ed.). Geneva. ISBN 978-2-9701028-5-4. Retrieved April 10, 2024 from https://sielok.hu/files/RM-world-report-2018.pdf

Weiss, G. (2001). Mountain forest policy in Austria: A historical policy analysis on regulating a natural resource. *Environment and History*, 335–355.

WKO, Fachverband der Seilbahnen. (2017). *Seilbahnfibel Winter 2006–2016.* Austrian Federal Economic Chamber's Association of Cable Cars. Vienna.

8 Sport Tourism and the Role of Open Spaces in Cities

Lawal M. Marafa and Minkun Bill Liu

Introduction

Historically and in contemporary times, tourism and sport are interrelated phenomena. While tourism attracts people traveling from one place to another, sports attract people for a specific reason and sometimes travelers will participate in or witness sports or events related to sports (Chalip, 2006; Gibson, 1998; Gibson et al., 2003). This relationship dates back to historic times and subsequently involved the early development of both tourism and sports as major industries.

As the industry developed, major sporting events were conducted in outdoor and open spaces as well as some indoor events. As this evolved, the open spaces changed and their relationship to sport and tourism changed significantly. While there are many sources in the literature that chronicle the evolution of tourism and leisure, historically, sport was one of the instigators of travel, as people visit major sport events to spectate or partake in them (Ritchie et al., 2009; Smith, 2009). For example, sport acted as an instigator to travel in early Greece and Rome. As sports evolved, people sought places that had open space, and indeed green space for certain sports activities and seek modified spaces and constructed facilities in more recent times. The attraction of open spaces and green spaces for sports activities has been well documented (Beedie, 2003; Chen, 2013; Weiss et al., 1998).

Specifically, after World War II, people were searching for leisure, entertainment, and for ways to improve their quality of life. Such leisure and entertainment they found in attending outdoor activities in which some are in the form of organized sports in the outdoors. Eventually, the content of outdoor activities changed with the introduction of extreme and adventure sports such as rock climbing, mountaineering, camping, river trekking, skiing, and more. As these outdoor sports became popular, there was an increase in participation worldwide. Some people travel to attend and participate while others view the trend as a way to approach health, leisure, and entertainment. Sports in this category include running, cycling, bird watching, fishing as well as jogging, marathons, and football (Chappelet & Lee, 2016; Zauhar, 2004). When discussing sport tourism in any perspective, it is important to identify and analyze resources related to sport tourism that could include both natural and man-made resources.

DOI: 10.4324/9781003476658-10

Identifying and analyzing such resources is important, given that sports tourism products or markets can be classified as either leisure or business (Chalip & Costa, 2005). For destinations to benefit from the opportunity that sports tourism provides, it is important to understand that leisure sports can be represented by events, cultural heritage, fantasy camps, sports cruising, outdoor and adventure travel, health spas, and more. On the other hand, business sport tourism is limited to sport organizations, competitive team sports, and individual sports. Any of these categories has the capacity to attract visitors to such destinations, some of whom will travel short, medium, or long haul to witness the occasion. As this trend grows and becomes highlighted by major sport events like the Olympics, FIFA World cup, Women's Tennis Association (WTA), Professional Golfers' Association (PGA) and more, many countries will incorporate sport tourism into their tourism calendar and market them accordingly. How these are utilized at different destinations will need further understanding.

Given that the majority of sporting events happen in the outdoors, some scholars (DeNizio & Hewitt, 2019; van Bottenburg & Salome, 2010) have lamented on the lack of long-term planning for sport tourism. Organizers and proponents in many cases focus on short-term planning, seeking to have success in the event. It is now beginning to appear that the local community must be involved to make an event successful. Such successes are now happening in many cities due to the number of spectators, visitors, and tourists that attend and bring enormous financial value to the destination.

The Relationship between Tourism and Sports

Sports and tourism are interrelated (Weed, 2009). While sporting events often bring people together, tourism also attracts a number of people to particular places. Leiper (1990, p. 370) had defined tourism as a "system comprising of three elements: a tourism or human element, a nucleus or central element and a marker or informative element". Markers form essential components of attractions, projecting meanings and signifying it is worth visiting. It is the markers that accord meaning and thus generate interest to tourists. In this context, sports attract participants, spectators, and many others constituting the human element. Markers represent the advertisements and promotions of the central element or the nucleus or indeed the place of where the sport is produced and consumed. To some people, the actual sporting event is the main attraction; to others, the personalities are the attraction while the sport venues are also attractive to some visitors (Ramshaw, 2011). There is also historical evidence that many sport areas were created in open spaces (DeNizio & Hewitt, 2019; Raitz, 1987; Wang et al., 2019) that were previously natural landscapes.

The connection between sports and open spaces or landscapes has been well documented (Manas, 2017; Ramshaw, 2011). For example, the Tour de France often results in the closure and re-direction of commuter traffic along the event's route. Although the Tour de France is synonymous with the rural landscapes of France, it now occurs in urban areas as in London 2007 (Balduck et al., 2011;

Palmer, 2010). Various marathons that are organized across the globe are often in cities like Boston, New York, London, Hong Kong, Nanning, and others. It is, therefore, necessary to identify the sports venues as landscapes (Ramshaw, 2011) as they are being created across the globe. Accepting the argument that sports is a form of culture (De Knop, 1990; Funk & Bruun, 2007), landscapes of sports are, therefore, cultural landscapes (Bale, 2003). While some landscapes remain unmodified (e.g., ski resorts, river rafting locations, mountaineering sites), some are specifically constructed and open for interpretation (e.g., golf). For example, a stadium can be seen as a park or green space *and* a site of sport/event production. Consequently, it is in such places that humans can develop their personal and collective identities.

These areas that are now major sports venues once supported natural and semi-natural habitats. In many urban areas across the globe, sporting grounds represent one category of open space that attracts attention and is exposed to visitation. Such sporting grounds are sometimes green, open spaces that contain parks, gardens, and other features that require ample space for leisure use. However, the relationship between sports and green spaces has been long and controversial. Although there is a strong relationship between sports and open spaces, the requirement for different sports needs to be evaluated.

Many people have been drawn to sports and sports venues even if they do not participate in the activities. This phenomenon started when many organized sports were established. People with interest thus become spectators generating fanfare to behold. When it was evident that organized sports were becoming important to cities, it attracted the expansion of venues and development of facilities to accommodate spectators and crowds, as depicted in the numerous stadia and sport complexes across the globe. While many spectators are locals, others travel further to attend and often spend a night or two.

In the second half of the 21st century, the facilities became commercialized. Cities also saw the advantage of such sporting events and their venues, particularly in subsequent branding. Some cities are branded as a result of the sports that they play and the venues that they have (Chalip & Costa, 2005; Chalip et al., 2003; Gibson, 1998; Vierhaus, 2018).

In articulating the role of sport tourism, it is necessary to emphasize the fact that all sports have distinctive profiles, both on the spaces where they are played and the public that they attract. Sport tourism also has the capacity to transform the societies that host such sports. There are four broad types of sports: organized amateur sports or team sports; international competitive sports; commercial sports; and informal, individual sports (Weed, 2009).

Relationship between Sports and Open Spaces

Historically, sports in general, and indeed major sporting events and festivals are a major cultural component of modern societies (Jonsson, 2003; Raitz, 1987). This cultural aspect of sport attracts a large crowd of people including visitors and tourists, creating a blend of locals, casual attendees, and fans who travel specifically to watch the event (Jones, 2008).

Although drawings found in pharaonic moments depicted various sporting activities believed to be played outdoors (Schnitzer et al., 2017; Zauhar, 2004), some other sports like the ancient game of Cuju (modern day football or soccer) were played in China from 206 BC specifically in open spaces (Guo & Li, 2017). In contemporary times, many major sport events like the Olympics, FIFA World Cup, and others depend largely on the outdoors and some built and modified environments to be successful (Kruger & Saayman, 2012).

Geographically, sporting events and the visitors that they attract transform spaces and the venues then become places where distinct activities occur (Tuan, 1977). If the space is demarcated and transformed into a sporting ground, it will be characterized by patterns, specific shapes, and sizes (Antrop, 2000). The sporting ground consequently provides a context for various experiences and social interactions that occur in societies (Bale, 2003; Raitz, 1987). The interactions that occur in these places and the grandiose nature of some of them depicts the cultural aspect of societies, thus generating various social experiences.

Sport tourism is one of the developing fields of international tourism involving travel, making it a significant segment of tourism and leisure sectors. According to scholars (Malchrowicz-Mosko & Munsters, 2018), whether or not sport tourism development takes the form of sport activities, it draws on local resources and consequently forms part of a complex dynamic of community life (Preuss, 2015). In this regard, sport tourists are increasingly interested in locations of important sport events. For example, Olympia, Greece, the site of the first Olympic Games in 776 BC, attracts contemporary tourists *en masse* (Yfantidou et al., 2017). Currently, sports fans and tourists travel extensively and enjoy visiting sport venues like stadia and in some cases also to sport museum (Fairley & Gammon, 2005; Malchrowicz-Mosko & Munsters, 2018), making such visits a very popular type of leisure and sport tourism. During and even after hosting major sporting events, infrastructure like stadia become epicenters for tourist and leisure events, establishing a positive and lasting legacy for the events (Gratton & Preuss, 2008). Examples of these infrastructures are the Bird's Nest in Beijing, China; Aspire Zone in Doha, Qatar, and Sports City in Dubai, UAE.

Globally, there is increased attention toward sport tourism. In the US, as earlier reported, sport (event) tourism generated an estimated $27 billion a year at the start of the millennium (Travel Industry Association of America, 2001). Since then, sport tourism as a special form of international tourism became increasingly popular. Consequently, sport fans, among who are tourists, became consumers alongside sport teams, engaging in sport-related experiences. The desire for sport experiences results in long distance travel to partake and witness events (Smith, 2007).

Around the world, many people travel significant distances to watch their favorite sports, sometimes on regular basis. These events include sports organized by the Association of Tennis Professionals (ATPs), PGAs, European Football leagues, etc. Major sports, sometimes referred to as hallmark events, can be referred to as "major fairs, expositions, cultural and sporting events of international status often held either on a regular or on a time to time basis" (Hall & Hodges, 1996, p. 17). This generates enormous economic impacts on the location and society, sometimes leaving behind a legacy (Hall & Hodges, 1996; Preuss, 2015). Some of these mega events are short lived and easy to manage, but they could still have long-term consequences for the community, both positive or negative (Gratton & Preuss, 2008; Preuss, 2015).

Sports, Leisure, and Green Spaces

The advent of modern sports generated interest in outdoor sports in which organized teams compete. Such organized sporting events have a long history in many parts of the world where the majority of events are conducted on grass/turf and involve hitting, kicking, catching, carrying, running, falling, scoring, and more (Beard, 2012). In many cases, the outdoor playing surfaces will be composed of varying types of surface cover and undulating landscapes. These might include flat surfaces (football, cricket, rugby, tennis, etc.), streams (rafting and eco-challenge), and sometimes undulating landscapes with vegetation ranging from trees to bare areas (Petrosillo et al., 2019).

Most sport participants need a well-manicured grass lawn to play on. In many communities and cities, open spaces and green spaces are found everywhere from town squares and parks to golf courses and elsewhere. Such places are often regarded as the green lungs of society and are accessible to thousands of people, including those who travel to attend sporting events. As people across the world continue to be drawn to outdoor sports, some cities have developed and promoted the indoorization of outdoor sports (van Bottenburg & Salome, 2010). Although certain sports take place in natural areas, select outdoor sports are now found indoors, including climbing, skiing, surfing, and others.

Tourists, Sports, Leisure, and Open Spaces

Although there is a strong relationship between sports, open spaces, and natural environments, there are various requirements for different sports and the sports generate differing interests. It is, therefore, incumbent that the relationship be investigated. Where the sport (event) attracts a large number of spectators, fans, and visitors (tourists), there is also the need for that phenomena to be understood to improve leisure experience activities.

Many sporting events drew fanfare and, in this context, the experiences of fans were socially determined giving the average person direct interaction with important sports people, places, and events (Jones, 2008). While many spectators are local folks, others travel to attend and often spend a night or two as tourists seeking to meet local people, see popular destinations, or partake in those events.

The season in which the sport event is held influences travel patterns, not only of the participants but the visitors, tourists, and indeed the diehard fans. In this regard, both the host of the event and the source of the visitors will be dependent on the season to facilitate travel and participation. While some seasons are desirable (e.g., snow for winter sports), some seasons and weather will be inclement (rains in outdoor sports like soccer, golf, tennis, etc.) for both participants and visitors.

As this phenomenon grows, it will be incumbent that organizers and destination managers devise strategies and tactics to circumvent any negative seasonal factors so that sport and tourism can flourish.

Given that many sports venues in contemporary times are man-made like stadia and playfields, destinations have the opportunity to establish these locations as popular spots for visitors. With the new generation of sports enthusiast and those that accompany them during group travel, the local stadium can act as a backdrop to a major visitor attraction. In addition to the sports edifice being a major attraction, it is also possible to make the venue a multi-use space for exhibitions, conferences, and conventions.

Case Examples of Sports and Tourism in Nanning and Hong Kong, China

Nanning

As a second-tier city located in South China, one that serves as the capital of Guangxi Province, Nanning prides itself with a lush, green environment and modern sport facilities. The city encourages local residents and visitors to take advantage of the facilities and events that they often host to indulge in sports both as active participants and fans. As in many healthy cities, Nanning prides itself with interesting adventure events like hiking in the forest, camping in the mountains, hanging by the cliffs, and more. Other activities in nature include swimming, river rafting, rappelling, or even getaway adventures near the city.

Nanning is known as the "Green City". It earned this sobriquet as it has abundant lush, subtropical foliage thanks to its humid subtropical climate, and surrounding hilly basin. Boasting with facilities, the government forges more sport events within the Association of Southeast Asian Nations (ASEAN) regional sports industry. In this regard, Guangxi has successfully organized a number of sports events with ASEAN that brought together a large number of visitors providing opportunities for cultural, economic, trade, and tourism activities. Table 8.1 provides some examples from the past few years.

At first, the city hosted the World Half Marathon Championships in 2010. Since then, it has been put on the sport and tourism map. The city of Nanning has also provided opportunities for cultural and educational escapades for young people, drawing more tourists to the area, specifically sports tourists. Consequently, Nanning showed strong national customs at the opening ceremony of the 45th World Gymnastics Championships. This occurrence was the first time that China had hosted an international sports event in a minority area of borderland that the province accommodated. As part of a series of major events in 2019, the China Cup International Football Championship was held in Nanning.

As a major outdoor sport destination, the city also hosted one of the most attractive circuits of the Union Cycliste Internationale (UCI) World Tour in 2019. At this event and in this context, the Nanning Qingxiu Mountain Circuit had many curves and large slopes, posing an interesting outdoor landscape to behold. Entering

Table 8.1 Mega Events in Nanning and Hong Kong

Place	Event	Type	Characteristics	Remarks
Nanning	World Gymnastics Championships	International	Gymnastics	Annual event
	China·ASEAN World Boxing Organization (WBO) International Championship	International	Boxing	World Boxing Organization
	China Cup International Football Championship	International	Football/Soccer	China Cup Football Carnival
	Sudirman Cup	International	World Mixed Badminton Championship	Every two years
	China-ASEAN International Marathon	International	Marathon	Alongside China-ASEAN Expo
	Union Cycliste Internationale (UCI)	International	Cycling	International Road Track Cycling
Hong Kong	The Hong Kong Rugby Sevens	International	Rugby	Annual rugby event
	Hong Kong Marathon	International	Marathon	Annual marathon event
	Hong Kong Open Golf	International	Golf	A PGA annual event
	International Dragon Boat Races	Cultural/ International	Boat Racing	Local Annual International event
	Longines Masters of Hong Kong, Equestrian	International	Equestrian	An indoor annual equestrian event

the Mashan Circuit, the speed was accompanied by passion, the green water ran through the green mountains, the idyllic scenery along the way was beautiful, and the rice fields and villages were scattered. The contestants cycled past the scenic area of Jinlun Cave and the tranquil scenery around Shanglong added to the enjoyment of "people riding in the painting". People along the way wore folk costumes, sang and danced to warmly welcome the athletes.

In general, Nanning organizes sports events and increases the difficulty of the games, showcasing national characteristics and making sports tourism more attractive. These techniques not only promote the development of sports, but also take the opportunity to show the scenery and culture of Nanning to the world, adding to the essence of sport tourism.

Hong Kong

The city of Hong Kong is located at the southern tip of China. As a city that boasts world class sport facilities, it attracts a variety of tourists, some of which travel to partake in sporting competitions or sport-like festivals that are ingrained in Hong

Kong culture. These sporting events range from PGA competitions, marathons to Dragon Boat Races (see Table 8.1). Hong Kong also uses its sporting events to capitalize on the natural environment and their numerous sports facilities to create more sporting events. The events are used to attract tourists and consequently generate economic benefits. The following two examples could be used to portray this assertion.

First is the annual Hong Kong Rugby Sevens series. This event generates a carnival-like atmosphere when it is hosted each year on a weekend in March. It is an annual event that was started in 1976, designed to promote sports in Hong Kong, and eventually it became a major tourism event for Hong Kong *and* the entire region.

As one of the biggest rugby events on the International Sporting Calendar, it attracts a large number of visitors to the city, resulting in the 40,000 seating capacity of the stadium being filled for all three days of the event. Another major event is the Hong Kong Tennis Open, sanctioned as part of the international series of the WTA. This event annually attracts more than 50 of the world's best female players and brings in quite a number of spectators from across the region.

While these two events are conducted in specifically modified open spaces, another major event that coincides with an ancient Chinese festival is the annual Hong Kong Dragon Boat Carnival. This is conducted along the coastlines and includes boat race participants from international as well as local teams. This is normally organized in a three-day period of intense racing filled with heart-pounding action, a profusion of color, the sounds of drums, and fans cheering for the participants. In this case, many local and visiting spectators line the harbor, transforming the space into a carnival-like atmosphere of modern sport and international party.

Sport Tourism and the Natural Environment

When discussing sport tourism in any perspective, it is important to identify and analyze resources related to sport tourism that could include both natural and man-made resources. Figure 8.1 provides a timeline of the development of sports in relation to natural open spaces. Most cities and communities use the development of such open spaces and sports to develop sport and tourism products that attract visitation. This is important given that sports tourism products or markets can be classified as either leisure or business (Boonsiritomachai & Phonthanukitithaworn, 2019; Chappelet & Lee, 2016). For destinations to benefit from the opportunity that sports tourism provides, it is important to understand that leisure sports can be represented by events, cultural heritage, fantasy camps, sports cruising, outdoor and adventure travel, and health spas among many others. On the other hand, business sport tourism is limited to sport organizations and competitive team and individual sports. Any of these categories have the capacity to attract visitors who will travel short, medium, or long haul to witness the events (Gibson et al., 2003; Jones, 2008).

As this trend grows and becomes highlighted by major sport events like the Olympics, FIFA World cup, WTA, PGA, and other regional sport events, many

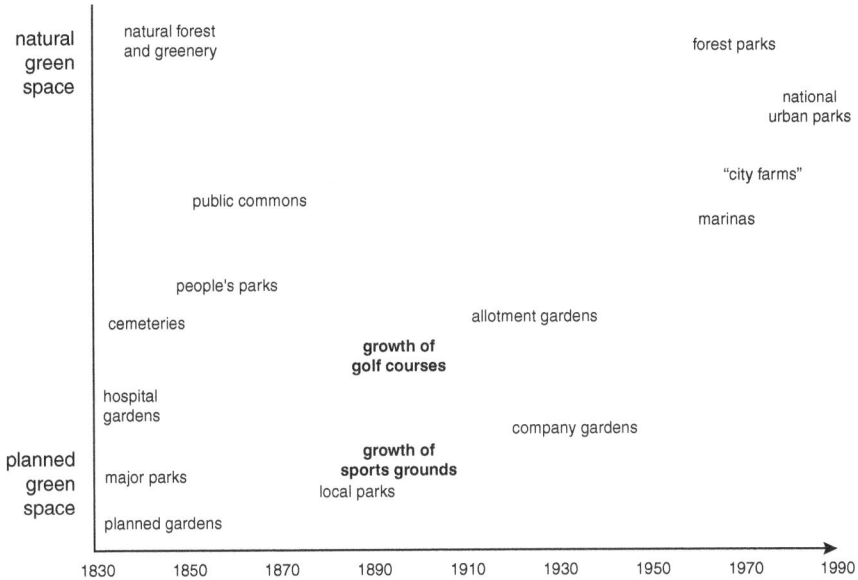

Figure 8.1 The Development of Green Spaces in Europe.

countries and corresponding destinations will incorporate sports tourism into their tourism calendar when major events are hosted and market them accordingly.

Given that majority sporting events happen in the outdoors, some scholars (Peric et al., 2018) have lamented on the lack of long-term planning for sport tourism. Organizers and proponents in many cases focus on short-term planning, seeking to have success in the event. It is now beginning to appear that the local community must be involved to make an event successful (Boonsiritomachai & Phonthanuki-tithaworn, 2019). Such successes are now being recorded regarding the number of spectators, visitors, and tourists that attend to track the enormous financial benefits they provide to the destination (Funk & Bruun, 2007).

Conclusion

While it is true that sport tourism is becoming buoyant, destination marketing officials and managers often have the task of highlighting the tourism component given that some of the spectators are visitors. Consequently, sport tourism is now featuring in promoting and marketing destinations where major sport events are held.

In the second half of the 21st century, sport facilities became commercialized. Cities also saw the advantage of such sports and their venues. Some cities are branded as a result of the sports that they host and the venues where they happen. As a component of tourist and visitor economy, there is existing difficulties in understanding theories of destination image and subsequent branding of the place and the sport event itself. This difficulty has also been identified in other aspects of tourism.

In order for the whole package of sport tourism to be successful, there is the need for the sport event to be consistent with community values to ensure there will be overwhelming community support for the event. Wherever there is community support, tourism flourishes and this will be the case with sport tourism. Such community involvement and support is seen in the previous examples in the cities of Nanning and Hong Kong as evidenced in this chapter. As it is evidenced in many studies including these ones, sports are highly dependent on the natural environment and are often influenced by seasonality. This assertion is also the case with tourism.

Investigating the relevance of sports tourism in a destination, scholars, policy makers, and proponents will have to understand the open spaces and the landscape of cities as well as the role of seasonality. For most open spaces in cities, conducting sport within them as events can increase the vitality of such locations, increasing the quality of life of the city, and making them tourist attractions. Where such sport events are conducted, it can also favor the enhancement and renovation of the open space in destinations, thus enhancing its tourism potential.

References

Antrop, M. (2000). Changing patterns in the urbanized countryside of Western Europe. *Landscape Ecology, 15*, 257–270. https://doi.org/10.1023/A:1008151109252.

Balduck, A. L., Maes, M., & Buelens, M. (2011). The social impact of the Tour de France: Comparisons of residents' pre- and post-event perceptions. *European Sport Management Quarterly, 11*(2), 91–113. https://doi.org/10.1080/16184742.2011.559134.

Bale, J. (2003). *Sports geographies.* Routledge.

Beard, J. B. (2012). *History of sports field turfgrass surfaces.* Sports Turf Manager. Retrieved from https://archive.lib.msu.edu/tic/stnew/article/2012win1a.pdf

Beedie, P. (2003). Mountain guiding and adventure tourism: Reflections on the choreography of the experience. *Leisure Studies, 22*(2), 146–167. https://doi.org/10.1080/026143603200068991.

Boonsiritomachai, W., & Phonthanukitithaworn, C. (2019). Residents' support for sports events tourism development in Beach City: The role of community's participation and tourism impacts. *SAGE Open, 9*(2). https://doi.org/10.1177/2158244019843417

Chalip, L. (2006). Towards social leverage of sport events. *Journal of Sport & Tourism, 11*(2), 109–127. https://doi.org/10.1080/14775080601155126.

Chalip, L. & Costa, C. A. (2005). Sport event tourism and the destination brand: Towards a general theory. *Sport in Society, 8*(2), 218–237. https://doi.org/10.1080/17430430500108579

Chalip, L., Green, B. C., & Hill, B. (2003). Effects of sport event media on destination image and intention to visit. *Journal of Sport Management, 17*(3), 214–234. https://doi.org/10.1123/jsm.17.3.214.

Chappelet, J. L., & Lee, K. H. (2016). The emerging concept of sport-event-hosting strategy: Definition and comparison. *Journal of Global Sport Management, 1*(1–2), 34–48. https://doi.org/10.1080/24704067.2016.1177354

Chen, X. J. (2013). Sports venues and outdoor landscape design. *Advanced Materials Research, 726–731*, 3600–3603. https://doi.org/10.4028/www.scientific.net/amr.726-731.3600.

De Knop, P. (1990). Sport for all and active tourism. *Journal of the World Leisure and Recreation Association,* Fall, 30–36. https://doi.org/10.1080/10261133.1990.10559120

DeNizio, J. E., & Hewitt, D. A. (2019). Infection from outdoor sporting events—More risk than we think? *Sports Medicine, 37*(5). https://doi.org/10.1186/s40798-019-0208-x

Fairley, S., & Gammon, S. (2005). Something lived, something learned: Nostalgia's expanding role in sport tourism. *Sport in Society, 8*(2), 182–197, https://doi.org/10.1080/17430430500102002

Funk, D., & Bruun, T. J. (2007). The role of socio-psychological and culture-education motives in marketing international sport tourism: A cross-cultural perspective. *Tourism Management, 28*(3), 806–819. https://doi.org/10.1016/j.tourman.2006.05.011

Gibson, H. J. (1998) Sport tourism: A critical analysis of research. *Sport Management Review, 1*(1), 45–76. https://doi.org/10.1016/S1441-3523(98)70099-3

Gibson, H., Willming, C., & Holdnak, A. (2003). Small-scale event sport tourism: Fans as tourists. *Tourism Management, 24*(2), 181–190. https://doi.org/10.1016/S0261-5177(02)00058-4

Gratton, C., & Preuss, H. (2008). Maximizing Olympic impacts by building up legacies. *The International Journal of the History of Sport, 25*(14), 1922–1938. https://doi.org/10.1080/09523360802439023

Guo, J. Q., & Li, R. (2017). The development of leisure sports in ancient China and its contemporary sports culture value. *Advances in Physical Education, 7*, 377–382. https://doi.org/10.4236/ape.2017.74031

Hall, C., & Hodges, J. (1996). The party's great, but what about the hangover? The housing and social impacts of mega-events with special reference to the 2000 Sydney Olympics. *Festival Management and Event Tourism, 4*, 13–20. https://doi.org/10.3727/106527096792232414

Jones, I. (2008). Sport fans and spectators as sport tourists. *Journal of Sport & Tourism, 13*(3), 161–164. https://doi.org/10.1080/14775080802327102.

Kruger, M., & Saayman, M. (2012). Creating a memorable spectator experience at the Two Oceans Marathon. *Journal of Sport & Tourism, 17*(1), 63–77. https://doi.org/10.1080/14775085.2012.662391

Leiper, N. (1990). Tourist attraction systems. *Annals of Tourism Research, 17*(3), 367–384. https://doi.org/10.1016/0160-7383(90)90004-B

Malchrowicz-Mosko, E., & Munsters, W. (2018). Sport tourism: A growth market considered from a cultural perspective. I do movement for culture. *Journal of Martial Arts Anthropology, 18*(4), 25–38. https://doi.org/10.14589/ido.18.4.4

Manas, A. (2017). A companion to sport and spectacle in Greek and Roman antiquity. *The International Journal of the History of Sport, 34*(16), 1801–1803. https:/doi.org/10.1080/09523367.2018.1463501

Palmer, C. (2010). 'We close towns for a living': Spatial transformation and the Tour de France. *Social & Cultural Geography, 11*(8), 865–881. https://doi.org/10.1080/14649365.2010.523841

Peric, M., Durkin, J., & Vitezic, V. (2018). Active event sport tourism experience: The role of the natural environment, safety and security in event business models. *International Journal of Sustainable Development and Planning, 13*(5), 758–772. https://doi.org/10.2495/SDP-V13-N5-758-772

Petrosillo, I., Valente, D., Pasimeni, M. R., Aretano, R., Semeraro, T., & Zurlini, G. (2019). Can a golf course support biodiversity and ecosystem services? The landscape context matter. *Landscape Ecology, 34*, 2213–2228. https://doi.org/10.1007/s10980-019-00885-w

Preuss, H. (2015). A framework for identifying the legacies of a mega sport event. *Leisure Studies, 34*(6), 643–664. https://doi.org/10.1080/02614367.2014.994552

Raitz, K. (1987). Commentary: Place, space and environment in America's. *Leisure Landscapes, 8*(1), 49–62. https://doi.org/10.1080/08873638709478497

Ramshaw, G. (2011). The construction of sport heritage attractions. *Journal of Tourism Consumption and Practice, 3*(1), 1–25. http://hdl.handle.net/10026.1/11568

Ritchie, B. W., Shipway, R., & Cleeve, B. (2009). Resident perceptions of mega-sporting events: A non-host city perspective of the 2012 London Olympic Games. *Journal of Sport & Tourism, 14*(2–3), 143–167. https://doi.org/10.1080/14775080902965108

Schnitzer, M., Schlemmer, P., & Kristiansen, E. (2017). Youth multisport events in Austria: Tourism strategy or just a coincidence? *Journal of Sport & Tourism, 21*(3), 179–199. https://doi.org/10.1080/14775085.2017.1300102

Smith, A. (2009). Theorising the relationship between major sport events and social sustainability. *Journal of Sport & Tourism, 14*(2–3), 109–120. https://doi.org/10.1080/14775080902965033

Smith, A. C. T. (2007). The travelling fan: Understanding the mechanisms of sport fan consumption in a sport tourism setting. *Journal of Sport Tourism, 12*(3), 155–181. https://doi.org/10.1080/14775080701736924

Travel Industry Association of America. (2001). Travel statistics and trends. Retrieved December 15, 2019 from www.tia.org.com

Tuan, Y. F. (1977). *Space and place.* Arnold.

Van Bottenburg, M., & Salome, L. (2010). The indoorisation of outdoor sports: An exploration of the rise of lifestyle sports in artificial settings. *Leisure Studies, 29*(29), 143–160. https://doi.org/10.1080/02614360903261479

Vierhaus, C. (2018). The international tourism effect of hosting the Olympic Games and the FIFA World Cup. *Tourism Economics, 25*(7), 1009–1028. https://doi.org/10.1177/1354816618814329

Wang, H., Dai, X., Wu, J., Wu, X., & Nie, X. (2019). Influence of urban green open space on residents' physical activity in China. *BMC Public Health, 19*, 1093. https://doi.org/10.1186/s12889-019-7416-7

Weed, M. (2009). Progress in sports tourism research? A meta-review and exploration of futures. *Tourism Management, 30*, 615–628. https://doi.org/10.1016/j.tourman.2009.02.002

Weiss, O., Norden, G., Hilschers, P., & Vanreusel, B. (1998). Ski tourism and environmental problems. *International Review for the Sociology of Sport, 33*(4), 367–379. https://doi.org/10.1177/101269098033004004

Yfantidou, G., Spyridopoulou, E., Kouthouris, C., Balaska, P., Matarazzo, M., & Costa, G. (2017). The future of sustainable tourism development for the Greek enterprises that provide sport tourism. *Tourism Economics, 23*(5), 1155–1162. https://doi.org/10.1177/1354816616686415

Zauhar, J. (2004). Historical perspectives of sports tourism. *Journal of Sport Tourism, 9*(1), 5–101. https://doi.org/10.1080/1477508042000179348

9 Managing Sport Tourism in Communities

Case Study in China

Lijun Jane Zhou and Miklos Banhidi

Introduction

Sport tourism is one of the fastest-growing market segments in the world (Hritz & Ross, 2010). Sport tourism products can be a large part of sales offers in destinations. There are well-known communities worldwide which focus on selling several sectors of sport industry. These products can be filled with rich content based on the surrounding environment. The development of environmental factors at destinations can offer better conditions for active tourism experiences. For this reason, tourists seek a good place to travel to refresh.

China's tourism industry has made it their target to become one of the world's favorite tourism destinations, so there is an intensive aspiration to develop new products, such a sport tourism. Due to the success of China's economy, tourism destinations have enough power to form their areas into attractive Sport Tourism Communities, where all sport industry subsections are presented and well-integrated within the tourism industry. There should be sport-effective indoor and outdoor areas, simple access to facilities, friendly tourism guides, and numerous programs and activities from which they can choose.

In this chapter sport tourism models were analyzed, and explained how these models can be integrated into tourism strategic development.

Communities and Sport Tourism around the World

It is necessary for tourism destinations to innovate in order to remain competitive in an increasingly global environment (Baggio & Cooper, 2010; Higham, 2005). According to tourism scholars, the relationship between tourism development, satisfaction, and importance of community dimensions is generally nonlinear with citizen involvement and public services (Allen et al., 1988). The interest in physical activity during vacations has indicated market growth of the sport tourism industry and has greatly modified strategies within the industry (Bouchet et al., 2004).

Sport tourism product providers offer more than just events and activities. Their products cause benefits and contribute to a healthier, fairer, greener, stronger, and wealthier living space around the destination. The success depends on the sustainable use of the natural environment, which should be kept clean and safe, without any extreme disturbances like light and noise pollution. These aesthetic and

DOI: 10.4324/9781003476658-11

emotionally stimulating surroundings promote feelings of fulfilment in whatever activity that is chosen by the individuals. For example, watching a sport event, or riding a bicycle that has been surfaced to enable challenging maneuvers without obstruction of motor vehicle traffic.

Tourism destinations facilitators frequently ask, what is an optimal environment to experience sport in the best way? Under which circumstances can tourists experience wellness while enjoying spectacles or experiencing the joy of movement?

All tourists are different, and, therefore, it is vital to understand their expectations. Tourism communities can be successful if they create a complex sport tourism model. By including all segments of the sport industry, this ensures that everyone can find an appropriate fit (Figure 9.1).

A sport tourism community model can be understood if it is an integral part of the community structure and if many stakeholders are involved. Dayton (1971) suggested it should be based on evidence that the growth and regulation of the component population in the community are affected in a predictable manner by natural and physical disturbances and by interaction with other species in the community.

If any community wants to target sport tourists, they should work on creating high quality surroundings for local citizens. After establishing tourism services and living zones, there is a large amount of pressure from the citizens to make their environment more attractive for visitors. One solution was to create directorates in city councils to amalgamate public sport interests into one consolidated department. The consolidation of sport activities in the public sector led to debates on the extent to which such activities should be a commercial or municipal enterprise (Hall & Page, 1999).

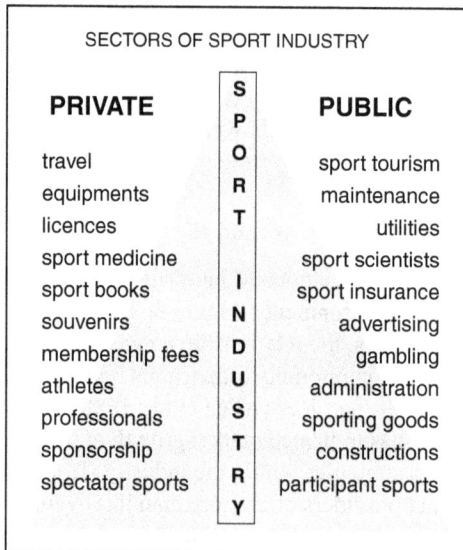

SECTORS OF SPORT INDUSTRY		
PRIVATE	S P O R T I N D U S T R Y	**PUBLIC**
travel		sport tourism
equipments		maintenance
licences		utilities
sport medicine		sport scientists
sport books		sport insurance
souvenirs		advertising
membership fees		gambling
athletes		administration
professionals		sporting goods
sponsorship		constructions
spectator sports		participant sports

Figure 9.1 Sectors of Sport Industry.

In tourism, the acceleration of urbanization processes is remarkable. Investors in urban areas became interested to take a part in tourism, connecting sport providers with one another. The majority of modern sport stadiums are owned by bigger cities, which gives them opportunities to host large-scale events. Although this is beneficial, urban areas present negative effects such as air, light, and noise pollution. Thus, it is understandable that some metropolitan cities like Beijing started the lower amount of carbon tourism (Cai & Wang, 2010).

Sporting event spectators are interested not only in the competition but also the advantages of tourism services in the host community. For example, the free capacity at the Wimbledon Tennis Centre is sold for tourists successfully. In contrast, rural regions do not have big facilities, but they can offer attractive open space for outdoor sports, and related activities tended to attract new residents and jobs.

The development of sport facilities was dominated by the changing expectations of tourists (Table 9.1). Travelers recognized the value of tourism activities and began to see past basic needs, seeking meaningfulness, relevance, and value in their travel experiences (Edginton et al., 2004a). Many of these expectations vary within different cultures, generations, and genders. In the present, a quick developing tourist from China often pays more attention to quality, including scenery and special services (Wang et al., 2020). Many of the sport tourists expect more green areas and wilderness at the destination, but other tourists want to avoid areas where they encounter the disadvantages of natural disturbances.

However, through tourism development, communities are trying to create attractions and services for the *tourist*'s interests. This means that communities no longer belong only to their own citizens. During high season in tourism regions, there are more tourists than local citizens, which influences the local development. Well-known and highly visited sport facilities offer leisure experiences for tourists as well. Many large-scale sport programs, such as sport festivals (Sofield & Sivan, 2003) and sport events (e.g., Tour de France, Formula 1, Olympics) are good reasons to mobilize millions of tourists every year. The question is, does tourism development support local leisure development, and is this development what citizens expect? The question is not easy to answer. Attractions help by hosting more tourists, which is good for the local economy. Local citizens can join those attractions and enjoy them, which can be enjoyable. On the other side, mass tourism can disturb the environment. Many times, tourists do not understand and follow the local community habits nor adhere to sustainable behavior.

Community Model Campaigns

In the last decades, different community projects have appeared to force the positive side of tourism development: UNESCO World Heritage Program, Family Friendly Workplace, Youth Friendly Community, For a More Livable Community, Flower cities, etc. The objective of these projects was to improve the quality of life of

Table 9.1 Expectations from Sport Tourism Offers

Wessinger and Bandalos (1995)	Driver et al. (1998)	Cordes (2003)	Pesonen Komppula (2010)	Jin et al. (2019)	Wang et al. (2020)
– Self determination – Competence – Commitment – Challenge	– Enjoy nature – Physical fitness – Rest – Reduce tension – Outdoor learning – Sharing similar values – Independence – Family Kinship – Introspection – Risk taking – Reducing	– Personal development – Social bounding – Therapeutic healing – Physical wellbeing – Stimulation – Freedom – Independence – Nostalgia	– Relax away from the ordinary – Escape from a busy everyday life – Have a hassle-free vacation – Get refreshed – Have a sense of comfort – Have an opportunity for physical rest	– Stimulus-avoidance – Intelligence seeking – Greater safety – Better service – Higher quality facilities	– Sceneries – Condition of slopes and snow – Training course feature – Cost

individual citizens and travelers through the creation of more livable communities. The criteria for the International Best Practices were that communities have:

- Landscape development;
- Heritage management;
- Sensitive environment practice;
- Sustainable groups; and
- Future plans – strategies.

The other famous project initiated by the World Health Organization is the Healthy Cities project, which brings cities who are committed to a comprehensive approach together. The purpose of the network is guided by a strong set of organizational principles. The project gives cities methodologies and structures that can be transferred from the health field into other fields that call for an integrated approach. The project is dealing with a lot of environmental and psychical factors, which are strongly connected to leisure areas. Also, ecological approaches in the communities became increasingly important, because besides improved life conditions, nature offers space and place for adventure, experience, sport, and relaxation. The citizens expect that these areas should stay clean, aesthetic, and effective for fun and for individual development.

Leisure in communities is altered by social issues. For example, it is often healthy and motivated people who are supported by a wide variety of leisure facilities and programs. Strong community structure and policies can secure safe and peaceful areas for leisure independent from natural dangers such as weather and pollution, or from negative social problems like crime. Thus, we should believe only in leisure areas that contribute to safe leisure environments. Communities should create platforms, where people can learn, be physically active, and socialize.

In a livable community, the basic structure is an intensive mix of activities, such as businesses, shops, entertainment, and housing, all located near public transportation. Leisure strategists in some international cities think similarly. Auckland, New Zealand, developed a residential Eight Zoning Program which ensures there is a sufficient population to support all leisure activities and that there is a high frequency of efficient public transportation around town centers in areas of change. This is where most of the increasing sport tourism population can be accommodated.

In Europe, sport tourism development has had positive changes. Campaigns help to manage some of their best tourism projects. For example, the Best Emerging European Rural Destinations of Excellence Award gave support to communities which:

- Offer an abundance of things to do and see;
- Maintain landscape based on preserved natural heritage;
- Establish handicrafts and specialty foods;
- Adopt a strategy of tourism development to establish sustainable economic development, and
- Support unchanged folk traditions, customs, and traditional crafts.

In one of our former surveys, the trials had to mention the favorite places of the local citizens (Banhidi, 2004). The answers were very interesting to us. It was much more important for the citizens to have meeting places mainly in outdoor areas, which offer a healthy and aesthetic environment. Most of the local citizens are not looking for the well-known attractions and they suggest them only for tourists. They prefer places where they can meet their own friends, rest, practice skills, and places that are not very expensive.

Environmental Conditions for Sport Tourism Development in Qingyuan County

China's tourism industry became the world's largest domestic tourist industry and the fourth largest international destination (Sofield & Li, 2011). The success was due to the re-birth of entrepreneurship and the market economy, structural changes (Sofield & Li, 2013), and supporting event sport tourism in China by hosting many large-scale events such as the Asian Games (Leung et al., 2014).

In July 2017, we had a site visit in Qingyuan County to give the administration guidelines for their tourism development. The city is located about an hour north of Guangzhou, most well known for its beautiful caves and waterfalls. The regional government proposed a strategic plan to build up competitive tourism products for international and Chinese visitors.

The local organizers' site visit focused on:

- Natural resources (mountain, water surfaces);
- Local traditions (small villages with unique architecture);
- Lifestyle of inhabitants, and
- Tourism services.

If a region would like to start tourism development, it should follow systematic guidelines. According to the sport tourism development model, the following elements should be taken into consideration:

- The environment is the background and the frame of the location, which should differ from tourist's home environment.
- Travel can be described in tourism as the move from home to a different destination. It can be accomplished by using different types of transportation, organized by a company or by the tourist themselves.
- In the last decade, tourists are seeking destinations where, beside good quality accommodation, they can find active programs like those they practice at home, or challenging ones to learn something new and to experience new sensations. This can be achieved through different sports and physical activity programs, which everyone knows.
- In tourism people don't want to compete, or necessarily increase physical abilities, they prefer more the cooperative sports, in which they can learn, have fun, and feel free to make mistakes.

We think that sport activities should only be offered to tourists if they are connected to the local culture and suit the natural and infrastructural conditions.

Observation of Geographical Resources

The Natural Environment

Natural resources are possibly the most important component of tourism. They serve aesthetically and help to create a healthy environment, but also as a challenging one, where visitors can experience unique activities. The natural environment includes climate, topographical relief, water surfaces, fauna, and flora. During our stay in July, the weather was very pleasant in Qingyuan and much more enjoyable in most areas of Southern China.

The fields in Qingyuan County have unique character (Figure 9.2). The rice fields, bamboo forests, and lotus flowers are not only pleasant to view, but they also tell a story about local lifestyles. The region also has rich water resources, so in the valleys, clean and fast-running rivers can be seen.

Social Environment

For tourism development, it is important to get the local community to support these endeavors, as tourism shouldn't be isolated from society. A successful tourism destination always involves the host community and their citizens, traditions, habits, institutions, and infrastructure. Former scientific studies reported that tourists' main expectation when traveling is to be able to "touch" the destination, to experience the best and unique services, preferred by locals. They would like to travel the same routes with local people to have authentic involvement. During our visit, we experienced outstanding hiking tours to the bamboo forests and the rice fields, which were attractive for tourist as well.

Figure 9.2 Rice Fields as Touristic Attraction in the County.
Source: Miklos Banhidi.

During our visit we experienced some of the following:

*   *Traditional architecture* – in this region, their artwork mirrored the art of China. Everywhere we looked, it was obvious that reconstruction had occurred, which make the area clean and organized looking, but we hope that the traditional image will not be lost. The most remarkable structures are the covered bridges (Figure 9.3), which functioned not just as river crossings, but also a place for socializing, selling goods, or just having fun. The history of the bridges can be better understood in the museum, but there are no guidelines on how to visit the other ones, or there is not any program around the bridges.
*   *The home-made food* – we tried local cuisine in the restaurants and we loved the local mushrooms, rice, and fresh fish from the river.
*   *Local art, handicraft* – straw hats, Chinese papercuts, bamboo products, etc.
*   *Traditional fashion, decorations, clothes* – on the streets, red lampions, placards, hangers, wall paintings, and more made for a unique atmosphere. We were missing typical folk costumes.
*   *Traditional theatre, games, music, and dance* – for China is very typical that citizens gather to play table or card games. When we visited the villages, we saw older adults playing those games and playing traditional music on the streets. We wished to join them.

The lifestyles of the people living in the countryside are connected to their daily work routines in the bamboo forests, rice fields, or mushroom tents. Tourists are often very eager to be a part of this lifestyle.

Figure 9.3 Covered Bridge in the Region.
Source: Miklos Banhidi.

Economic Environment

Rapid and far-reaching development transition has triggered corresponding restructuring in rural China, especially in the last two decades (Long et al., 2011). The fast-growing economy allows governing bodies to finance investments in the tourism sector, leading to visible positive changes in Qingyuan County. The region's famous economy is agriculture, but there is also industrial power, which can play an important role in tourism development. During our visit we were reassured that companies are interested.

Although the prices of most local services are reasonable, we found out that some prices are much higher. For example, a private hotel room costs as much as it does in Switzerland. There is also an entrance fee to use some hiking trails and tourism packages could decrease their costs.

Infrastructural Environment

Two types of infrastructure exist in tourism: static and dynamic. The static is global infrastructure, which is created by the country, region, or community, such as road systems, transportation, health care, and security institutions. The dynamic ones are established for tourism, such as hiking and bike trails, lookout towers, and swimming areas.

The road system in China is good, but there aren't always good transportation opportunities from the airports for international visitors. When visiting, there can be found many interesting dynamic facilities, which can serve as touristic attractions. In the city, a modern, multifunctional sport complex was built, which can host several sport competitions, although there are still no plans to sell it in tourism. There was a nice lake and a pool for swimming as well as nice hiking and biking trails. Also, they have pleasant hotel rooms and private accommodation for tourists.

With the new glass bridge that opened in June 2018, and some unique wooden waterfront buildings in Qingyuan (Figure 9.4) make for a good day trip from major cities like Shanghai, Guangzhou, and Hong Kong.

Experiencing Activities in Qingyuan

International tourism specialists report that, nowadays, tourist give priority to the quality of the programs around the destination (Kim et al., 2013). There can be interesting, planned sport activities for the visitors, as well as:

- Night walks in the old part of the village;
- Excursions to neighboring villages;
- Hiking (walking) tours to the covered bridges; and
- Hiking in the bamboo forest, rice fields, etc.

Conclusion

Visitors who travel to Qingyuan County will find unique conditions for their activities. From the observation of a SWOT matrix (Table 9.2), visitors and local hosts can prioritize expectations and decide what opportunities can be realized to develop sport tourism.

Figure 9.4 Hiking Trail along the Mountain Stream.
Source: Miklos Banhidi.

Summary

In tourism destinations, community group approaches can secure the expectations of the tourists, making it a less expensive option to other assessment approaches (Edginton et al., 2004). For sport tourists, the main expectation is to experience the sport culture, which is offered as a cooperation between the sport and tourism industry.

If the communities are interested in sport tourism, they should establish attractive, sports-friendly conditions for the visitors. For this to occur, they should develop local sport tourism strategies based on their environmental resources. In communities, there should be a harmony among the environmental (natural, socioeconomic, and infrastructure) factors and human activities which depend greatly on the social well-being and the creative coexistence of many cultures.

In the observed region of China, Qingyuan, tourism leadership could create a unique destination because of its hospitable people, its landscape, and the unique mix of activities and services. We suggest that host communities should compile sport tourism packages involving the local resources and market them with other tourism sectors. The region doesn't yet have enough experiences in the sports industry. Even though they have enough modern facilities, they should also provide other sectors, such as sport tourism professionals, more developed sport retailers, or better sports health care.

Table 9.2 Environmental Analysis of Qingyuan County

Strengths	Weaknesses
Natural	*Natural*
– Optimal weather,	– It is difficult to sustain
– Rich, scenic mountains and water surfaces	– The climate and space limit the development
– **Unique fields (bamboo, rice)**	*Social*
Social	– Limited information for international tourists
– Kind population	– Few people speak foreign languages
– Interesting traditions	*Economical*
– Typical Chinese lifestyle	– The event is not known yet for sponsors
– **Exciting excursions**	– Some accommodations are expensive
Economical	– Uncertain governmental support
– Strong Chinese economy	– Some services are expensive
– Cheaper and more frequent international airfares	*Infrastructural*
Infrastructural	– Difficult access from bigger cities
– Traditional architecture	– The hiking trail are not in a good shape everywhere
– Good quality highways	
– Developing hiking and biking trails, lookout towers	
Opportunities	Threats
Natural	*Natural*
– Building a botanic park	– Climate change
– To build deeper water surface for swimming	– The changes in the nature can cause unexpected results
– To establish a small animal garden	*Social*
Social	– Modernization can kill traditions
– More traditional elements can be managed	– Lifestyle of tourists will change
– Tourism information can be developed in other languages	– Competition with other events
– Internet sales can be developed	*Economical*
– More programs	– Lack of governmental support
Economical	– Sponsors won't join
– Tourism packages with different price categories	– Quick development increases the prices
– More attractive programs to attract governmental and non-governmental supporters	*Infrastructural*
Infrastructural	– Modern reconstructions can change the traditional landscape
– Better transportation system from airport with railway stations	

For sport tourism development in this region, many stakeholders play an important role. The best ways to encourage sport tourism development are to:

• Establish a network of sport tourism providers in the community;
• Offer tax benefits for sport associations to share their know-how with tourists;
• Give preference to local sports development;

- Come up with opportunities for tourists to play sports in public spaces and municipal sports facilities, and
- Encourage Chinese communities to join the international tourism networks.

References

Allen, L. R., Long, P. T., Perdue, R. R., & Kieselbach S. (1988). The impact of tourism development on residents' perceptions of community life. *Journal of Travel Research, 27*(1), 16–21.

Baggio, R., & Cooper, C. (2010). Knowledge transfer in a tourism destination: The effects of a network structure. *The Service Industries Journal, 30*(10), 1757–1771.

Banhidi, M. (2004). Promoting wellness: European perspective. In M. K. Chin, L. Hensely, P. Cote, & S. H. Chen (Eds.), *Global perspectives I integration of physical activity, sport, dance and exercise science in physical education. From theory to practice* (pp. 133–144). Hong Kong: Dept. of Physical Education and Sports Science, The Hong Kong Institute of Education.

Bouchet, P., Lebrun, A. M., & Auvergne, S. (2004). Sport tourism consumer experiences: A comprehensive model. *Journal of Sport & Tourism, 9*(2), 127–140.

Cai, M., & Wang, Y. M. (2010). Low-carbon tourism: A new mode of tourism development. *Tourism Tribune, 1*(5), 13–17.

Cordes, K. A. (2003). *Applications in recreation & leisure: For today and the future.* McGraw-Hill.

Dayton, P. K. (1971). Competition, disturbance, and community organization: The provision and subsequent utilization of space in a rocky intertidal community. *Ecological Monographs, 41*(4), 351–389.

Driver, B. L., Brown, P. J., & Peterson, G. L. (1998). *Outdoor recreation: A reader for congress – government.* Printing Office.

Edginton, C. R., Hudson, S., Dieser, R., & Edginton, S. (2004a). *Leisure programming.* McGraw-Hill.

Edginton, C. R., Jordan, D. J., DeGraaf, D. J., & Edginton, S. R. (2004b). *Leisure and life satisfaction: Foundational perspectives.* McGraw-Hill.

Hall, C. M., & Page, S. J. (1999). *The geography of tourism and recreation.* Routledge.

Higham, J. E. (Ed.). (2005). *Sport tourism destinations: Issues, opportunities and analysis.* Routledge.

Hritz, N., & Ross, C. (2010). The perceived impacts of sport tourism: An urban host community perspective. *Journal of Sport Management, 24*(2), 119–138.

Jin, X., Xiang, Y., Weber, K., & Liu, Y. (2019). Motivation and involvement in adventure tourism activities: A Chinese tourists' perspective. *Asia Pacific Journal of Tourism Research, 24*(11), 1066–1078.

Kim, K., Uysal, M., & Sirgy, M. J. (2013). How does tourism in a community impact the quality of life of community residents? *Tourism management, 36*, 527–540.

Leung, D., Li, G., Fong, L. H. N., Law, R., & Lo, A. (2014). Current state of China tourism research. *Current Issues in Tourism, 17*(8), 679–704.

Long, H., Zou, J., Pykett, J., & Li, Y. (2011). Analysis of rural transformation development in China since the turn of the new millennium. *Applied Geography, 31*(3), 1094–1105.

Pesonen, J., & Komppula, R. (2010). Rural wellbeing tourism: Motivations and expectations. *Journal of Hospitality and Tourism Management, 17*(1), 150–157.

Sofield, T. H., & Li, S. (2011). Tourism governance and sustainable national development in China: A macro-level synthesis. *Journal of Sustainable Tourism, 19*(4–5), 501–534.

Sofield, T., & Li, S. (2013). Tourism governance and sustainable national development in China: A macro-level synthesis. In B. Bramwell & B. Lane (Eds.), *Tourism Governance* (pp. 91–124). Routledge.

Sofield, T. H., & Sivan, A. (2003). From cultural festival to international sport- The Hong Kong dragon boat races. *Journal of Sport Tourism, 8*(1), 9–20.

Wang, J., Huang, X., Gong, Z., & Cao, K. (2020). Dynamic assessment of tourism carrying capacity and its impacts on tourism economic growth in urban tourism destinations in China. *Journal of Destination Marketing & Management, 15*, 100383. https://doi.org/10.1016/j.jdmm.2019.100383

Weissinger, E., & Bandalos, D. L. (1995). Development, reliability and validity of a scale to measure intrinsic motivation in leisure. *Journal of Leisure Research, 27*(4), 379–400.

Part III

Sport Event and Activity Management

Photo: Farhad Moghimehfar

10 Cycling Tourism Events

Motivations and Participation

Marcel Grooten and Lenia Marques

Introduction

In the leisure field, the amount of interest in cycling has seen a recent rise in different ways. Along with the individual willingness to participate in a cycling activity, which can be motivated by different factors, there has also been an increase in the interest related to different types of cycling events. Within sports tourism management, cycling in general, and cycling events in particular, play an important role for the regions in which they take place. They are important not only during the event itself, with all the immediate impacts, but they are also important in the moments preceding and following the event. Managing such an event, therefore, goes far beyond the organization of the occasion and must be considered in different dimensions throughout the process to maximize the positive impacts (economic, social, cultural, recreational, health, quality of life, among others) and minimize negative impacts (noise and overcrowding, among others).

In order to manage a sporting event properly, including cycling events, it is important to understand first, what motivates people to want to participate in these events.

This chapter provides insight on the motivations and participation surrounding cycling events by analyzing data collected in 2016 in two major events: Tour de France and Tour Down Under (Australia). This research links these factors with the current use of bicycles and the personal involvement in cycling.

Understanding motivations and participation will allow for better managerial practices. Therefore, this chapter contributes to understanding the socio-cultural impacts of sport events, a topic that is gaining more and more attention in the decision-making process of bidding for events. The results of this research provide insights into visitor behaviors and profiles, which can be useful information for stakeholders involved in organizing cycling events.

Cycling Events

Research has revealed that visiting hallmark sporting events has a positive impact on the active involvement of people in cycling activities (Berridge, 2012b; Derom et al., 2015; Derom & Ramshaw, 2016). However, the factors which are perceived as important for further involvement in sporting activities have yet to be explored regarding cycling events.

DOI: 10.4324/9781003476658-13

Two important themes emerge from the studies discussed so far about the organization and facilitation of (mass) community cycling events. Firstly, there is a large volume of published studies describing the individual increase in cycling participant involvement (Adrian et al., 2006; Davies et al., 2011; Funk et al., 2011; Halpenny et al., 2016; Kulczycki & Halpenny, 2014). Besides that, other studies emphasize that facilitating local cycling initiatives will have a positive impact on the attendance rates of professional cycling events (Deenihan & Caulfield, 2015; Mackellar & Jamieson, 2015). Understanding the motivations of cycling participants provides strategic input for the design and promotion of successful cycling events and limits amotivation when participants do not see or feel a reason to continue anymore (Pelletier et al., 1995). By promoting cycling events and investing in cycling infrastructure, governments all over the world are actively implementing policies with the intention to motivate residents, workers, and tourists to become more active (Adrian et al., 2006; Berridge, 2012a; Deenihan & Caulfield, 2015; Fullagar & Pavlidis, 2012; Jamieson, 2014; Mackellar & Jamieson, 2015; Ritchie et al., 2009; Rose & Marfurt, 2007). Estimations in the UK alone reported that "for every £1 of public money spent [on cycling], the funded schemes provide £5.50 worth of social benefit" (Barratt, 2017).

Cycling is an environmental and efficient form of transport for all ages. If governments want to encourage and engage new potential riders, it is imperative to remove crucial barriers by improving road and infrastructure conditions and create an inviting cycling environment (Rowe et al., 2016). By enhancing infrastructural networks, governments play an essential role in the facilitation of outdoor leisure activities. To date, several studies have demonstrated that the environmentally friendly appeal of cycling has a positive impact on cycling participation (du Preez & Heath, 2016; Meschik, 2012; Sun et al., 2017).

Organizing a successful mass event, such as a cycling tour, requires detailed knowledge: what are the motives for participation (Alexandris et al., 2012; Funk et al., 2011) and what are the prerequisites for creating an inspiring environment for cyclists (Sun, 2017; Sun et al., 2017). It is vital for governments to play a leading role in the marketing, publication, mapping, and navigation of their cycle route networks (Chatterjee et al., 2013). Many governments promote local traffic information and signage before the event takes place (Chatterjee et al., 2013), impose speed limits and legislate the mandatory distance between cars and cyclists on public roads (Garrard et al., 2008), and build secure facilities, such as parking areas (Barratt, 2017; Wardman et al., 2007). Governments can stimulate cycling involvement even more by promoting existing and new safe cycling paths (Zander et al., 2013) and urban planning (Rowe et al., 2016).

Within this context, it is important to understand why people are joining cycling events, either as active cyclists or as part of an audience.

Cycling Event Motivations

As with many leisure activities, particularly in sports, there are many reasons why people participate in cycling events. With initiatives such as the Olympics, where the idea of legacy has been increasingly highlighted, the connections between passive and active participation are also gaining more attention from scholars and policymakers.

In previous studies, there were three important motivations in which we will explore more in-depth in this section: well-being and health, tourism, and landscape, and challenge oneself.

According to some studies, when looking at the social dimensions of cycling event motivations, cycling is still being stigmatized as a subculture, a barrier which is rooted in cycling environments (Aldred & Jungnickel, 2014). In their research, Navarro et al. (2013) conclude that almost 50% of the cyclists who got involved in cycling for the first time were influenced by partners and friends, something which can be explained by the social character of cycling together.

Previous research (Adrian et al., 2006) revealed that first-timers and novice participants in an organized mass cycling event were significantly more active in the month after their participation. Well-documented research shows that meeting friends and families plays a central role in cycling event participation (Brown et al., 2009; Fullagar & Pavlidis, 2012; King et al., 2018; Schulenkorf, 2009; Smith & Stewart, 2007). Cycling together has a positive impact on the development of protective skills and perceiving cycling as a social activity builds confidence and leads to better route knowledge (Garrard et al., 2008; Rowe et al., 2016; Zander et al., 2013).

Women and older adults are two groups of cyclists who require extra support (Rowe et al., 2016). Women's cycling participation can be encouraged by supporting their skill-level development and by creating social support networks, with relevance to increase their confidence (Rowe et al., 2016; Spotswood et al., 2015). After participating in cycling events, female cyclists felt more competent when they ride their bikes (Fullagar & Pavlidis, 2012).

Herman (2015) and Van Cauwenberg et al. (2012) argue that there is a stable increase in motivation when the cyclist becomes older, as long their environment is safe and enjoyable. Compared with younger cyclists, older cyclists are also more interested in learning new things (Gibson & Chang, 2012). The development of cycling skills never stops; there is always the next level of engagement and a next level of feeling competent (Herman, 2015). Based on repetition, cyclists set goals, become more confident, and consequently more competitive, beginning to 'act' like cyclists and commit to specific rules and behaviors (Bonham & Lumb, 2012; O'Connor & Brown, 2007; Rejón-Guardia et al., 2017; Ritchie et al., 2010; Spinney, 2006).

Together with these social dimensions of a cycling event, it is also important to consider that people are traveling to and from the event. This means that the event itself can serve as a major attraction in a location which may not be visited otherwise.

Regardless of the distance from home to the tour, the time of year, or the length of the tour itself, participating in a cycling event has a touristic appeal. A number of studies have postulated a convergence between the touristic environment of a leisure activity and the motivations of the participants (du Preez & Heath, 2016; Gibson & Chang, 2012; Halpenny, 2010; Halpenny et al., 2016; Hinch & Holt, 2017;

Hungenberg et al., 2016; Kaplanidou & Vogt, 2007; King et al., 2018; Kulczycki & Halpenny, 2014; Moon et al., 2013; Reitsamer et al., 2016).

Two important motivational aspects related to cycling in general, and affecting cycling events in particular, is the gear involved as well as a sense of (self-) challenge. There are multiple motivational reasons why (new) gear and equipment encourage cyclist to participate in cycling. Digital apps, for example, allow cyclists to log their rides in detail which provides insight on their expected physical health outcomes, related, among others, with well-being (Lonsdale et al., 2014), the inclusion of body- and muscle-focused motives (Rocha & Gratao, 2018), and references to 'lifestyle' diseases, weight control, and the prevention of injuries (Derom et al., 2015). Modern equipment, and the skills to control it, provide cyclists with the opportunity of becoming members of a social network (Barratt, 2017; Griffin & Jiao, 2015; Sun, 2017), and provides cyclists the possibility to share their performances, compare them with others, create a sportive identity, and even become a role model for others (Griffin & Jiao, 2015; Smith & Treem, 2017; Stragier et al., 2015; Sun et al., 2017; Ward, 2014; West, 2015). More specifically, the rationale behind all this is what Spinney (2006, p. 727) explains as "the feeling of the kinesthetic activity in the body and what each rider sees and feels in proportion to their level of skill dealing with the activity of cycling in an environment such as the high mountains". Although this specific feeling cannot be 'captured' (Spinney, 2006), it is one of the most important motivations for cyclists. Depending on how much actual or expected pleasure cycling causes or the social rewards it brings, participants make psychological choices based on needs for health and well-being (Chatterjee et al., 2013) to optimise a favorable social impression (Rowe et al., 2016), to avoid undesirable impressions, and to enjoy the notion of others thinking of them as cyclists (O'Connor & Brown, 2007).

Over the years, different models have been used in order to measure sport motivations. These models include: the Sport Motivation Scale (SMS), developed by Pelletier et al. (1995); the revised Sport Motivation Scale (SMS-II) by Pelletier et al. (2013); the Motivation for Physical Activity Measure (MPAM) (Ryan et al., 1997); the Behavioral Regulation in Sport Questionnaire (BRSQ) (Lonsdale et al., 2008); the Recreational Exercise Motivation Measurement (RFMM) (Koruç, 2015); and a multidimensional model of motivation (Alexandris et al., 2012). These different models for measuring sport motivations have at least one thing in common: the theoretical foundation for measuring and understanding that individual sport motivation is based on the self-determination theory (SDT), developed by Ryan & Deci (1985). In SDT, the motivations for activating human behavior are mainly psychological. However, socio-ecological theory offers a more holistic approach that looks further than intra-personal motivations for behavior, and encompasses the social, physical, and policy environment (Brown et al., 2009; Rowe et al., 2016; Van Cauwenberg et al., 2012).

Social ecological theory serves as a guide that is frequently used for the development of "interventions to encourage physically active behaviour" (Rowe et al., 2016, p. 419) and is used in various studies concerning the increase of physical activity (Brown et al., 2009; Derom et al., 2015; Derom & Van Wynsberghe, 2015; Rowe et al., 2016). Policies concerned with encouraging physical behavior are the organization of participatory cycle tours and the investment in cycling infrastructure.

For this research, the Cyclist Motivation Instrument (CMI), developed by Brown et al. (2009), is adopted. The CMI is a model that measures cycling motivations with the social ecological theory as the foundation. This model is relevant for this research because it encompasses the social, physical, and policy environment. The CMI consists of five dimensions: social, embodiment, self-presentation, exploring environments, and physical health outcomes.

In order to analyze these dimensions of motivation within the larger framework of participation in cycling events, it is important to mention the study of McIntyre (1989), where personal meaning of participation is conceptualized as 'enduring involvement' (EI). EI is derived from "a review of relevant recreation and consumer behavior literature" (McIntyre, 1989, p. 167), and is measured following a multidimensional approach using the three factors of self-expression, centrality, and attraction (McIntyre, 1989). Later on, this scale was updated by Kyle et al. (2007), who created the Modified Involvement Scale (MIS), which included social bonding.

This chapter will focus on the demand side to understand the factors of motivation in relation to participation to provide insights on improving cycling tourism management.

Methodology

This study is part of a larger research project about cycling events. The questionnaire was composed of different parts related to motivation, participation, and quality of life. The questionnaire was translated and adapted to different cycling events in different countries. Both the CMI and MIS were used as a basis for the questionnaire. The reliability of the 12-item scale for motives to participate in watching the tour was tested and obtained Cronbach $\alpha = 0.861$.

For this study, a clipping of the data was made, with a focus on two different countries, Australia and France, on socio-demographic variables to draw a profile, as well as variables related to motivation and participation. The sample consists of 683 respondents. However, to better understand their profile and motivations, we focused the analysis on a selection of 414 respondents who mentioned they had participated in an organized cycling tour in the 12 months prior to the data collection. We have, therefore, selected 295 respondents (76%) from an original database of 388 visitors who visited the Australian Tour Down Under (TDU) cycling event in January 2016, and selected 121 respondents (51%) from an original 235 visitors who visited the Tour de France (TDF) in July 2016. SPSS 25 was used for data analysis.

Findings and Discussion

In many sports and the discourses surrounding them, including in policy making, there is often the assumption that hosting a major sport event has impact on physical activity, well-being, and general quality of life. Major sport events, such as the

Olympics, are increasingly focusing on long-term legacy. This is also a way of justifying such an investment of what is, more often than not, the contributions of taxpayers.

The connection between hosting sports and the practice of sports is a link that is sometimes too readily made. Therefore, to try and understand this relationship, we analyzed the motivation scale between different groups of respondents. An Independent Samples Test with Levene's Test for Equality of Variances (equal variances assumed) was performed to compare means between the respondents who joined and those who did not join in the 12 months before the data collection. Out of the scale's 12 items (listed in Table 10.1), all items except two presented a significant difference ($p > 0.05$). The two exceptions were "to use my bike gear / equipment" ($p = 0.56$) and "to share skills and knowledge" ($p = 0.84$).

These results show that there is a relationship between the motivations to attend the tour and the fact that respondents have joined a tour previously. For the respondents who did join a tour, and who are, therefore, more active, most motivations are significantly higher. They are, therefore, more motivated to attend a major cycling event and their level of participation and engagement is higher.

Cycling is an important physical activity for many people and because of this, studies have focused on different aspects of cycling, including motivation. Findings of this study focus on cycling spectators of two tours (TDU in Australia and TDF in France) who have participated themselves in a cycling tour in the previous 12 months before the survey ($n = 414$).

A comparison of the respondents from the two countries revealed that 79% of the visitors in Australia are male versus 74% in the TDF. Although the majority of respondents are male, female presence is still relevant (see Figure 10.1). If we consider this sport has become increasingly popular among women, this fact is important, as women are a growing segment for which needs should be accounted for, especially in terms of confidence within networks (Rowe et al., 2016; Spotswood et al., 2015).

In terms of accommodation, 24% of the visitors in Australia and 20% of the visitors in France sleep at home which means that nearly a quarter of the event visitors do not have to travel far and do not require accommodation. Although

Table 10.1 Scale Items

1	For adventure
2	To gain experience
3	To keep physically fit
4	To meet new people
5	To experience elite cyclists' performance
6	To see other parts of the country / world
7	For a challenge
8	To use my bike gear / equipment
9	To show myself that I can do it
10	For relaxation
11	To share skills and knowledge
12	To be with friends and family

Gender

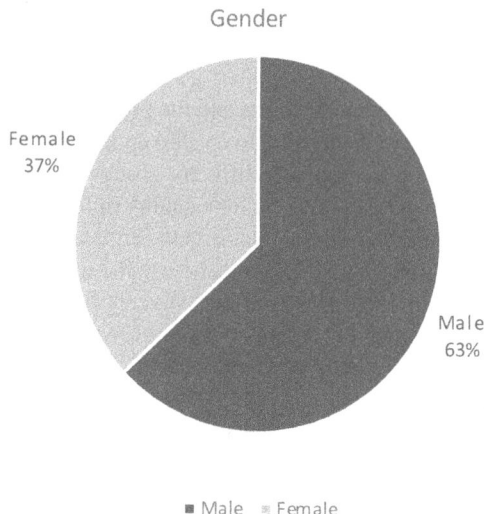

■ Male ▨ Female

Figure 10.1 Gender Prevalence for Both Tour Down under (Australia) and Tour de France.

these results are similar for accommodation, they vary in terms of nationality, being that 96% of the TDU visitors is Australian and only 45% of the visitors in the TDF is French. The TDF is, therefore, much more international in terms of its visitors. This can be explained firstly by the location itself, as Australia is more difficult to get to, being an island, and France is already a successful tourist destination, with the TDF being a part of the country's touristic products portfolio.

It is interesting to note that for both cycling events, 77% of the visitors have visited the event before. This suggests that event attendees are very likely to be repeated visitors. Given this fact, it is important to understand their motivations.

As it can be observed from Figure 10.2, the means in France are in general lower than in Australia. This difference might be explained by the fact that people expect more from that tour, or simply that Australians tend to score their perception higher, since most of the respondents from TDU are from Australia. This is also reflected in the statistically significant differences between the two countries in 11 of the 12 items describing the motives for participation in an organized tour. The only exception to this (item *not significantly* different) is the 'for adventure' (M = 3.64 in Australia and M = 3.44 in France). The highest score for both countries was on the item 'keep physically fit' (M = 4.36 in TDU and M = 3.89 in TDF), followed closely by 'to be with family and friends' (M = 4.11 in TDU and M = 3.49 in TDF).

There are, therefore, four main motives for respondents to attend these cycling events, which relate to (1) health and well-being, (2) social bonding, (3) challenge, and (4) relaxation.

Respondents seem to consider that elite cyclist performance is less important. This highlights the fact that, although this does play a role, personal well-being and

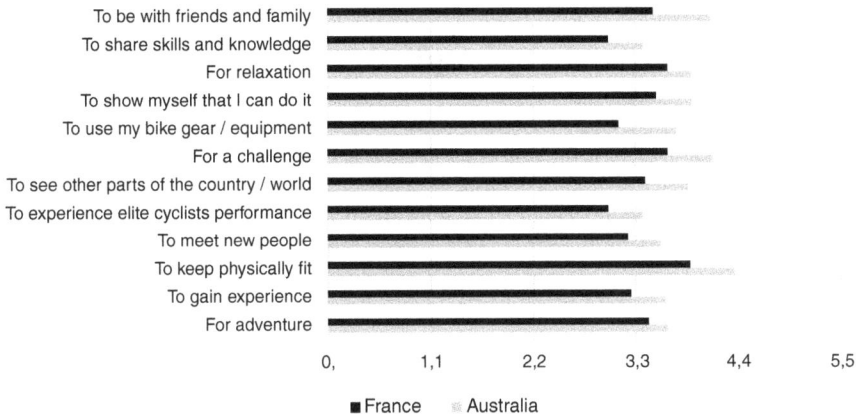

Figure 10.2 Motives to Participate (Means Based on a 5-Point Likert Scale).

social moments are much more important. Respondents use the event as a catalyst to provide opportunities for well-being (being fit or relaxation, or feeling good about taking on a challenge), and the event also functions as a platform for social interaction. The fact that only 8% of respondents in the TDF and 18% in the TDU were traveling alone, emphasizes the importance of the social dimension in such an event. This corroborates previous studies pointing out for the importance of the cycling event as a social activity (e.g. Brown et al., 2009; Fullagar & Pavlidis, 2012; King et al., 2018; Schulenkorf, 2009; Smith & Stewart, 2007).

The motives for participation also show some significant differences between the two countries when compared with the independent variables of gender and age. Australian visitors differ only on two items between male and female participants. Females significantly attach more value to 'gain experience' (t120 = −2.76, p < 0,01) and the item 'show myself I can do it' (t118 = −2.14, p < 0.05). This goes in line with the increased presence of female in cycling events (Rowe et al., 2016), and this is an aspect that event managers should consider when organizing the event in certain destinations.

Dividing visitors into two groups based on age (median age TDU = 47 years), there are five items with significant differences. People younger than 47 have higher scores on the items 'for adventure' and 'to gain experience', while people age 47 and over have higher scores on the items 'to keep physically fit', 'for relaxation', and 'to be with friends'. Male and female visitors of the Tour de France significantly differ on the item 'be with friends and family'; M = 3.88 for females and M = 3.35 for males (t119 = −2.25, p < 0.05). Dividing visitors into two groups based on age (median age TDF = 42 years) shows that there are three items where the younger group scores significantly higher; 'to gain experience', 'to experience elite cyclists' performances', and 'for a challenge'. This suggests that destinations and event managers should consider that younger groups should be facilitated with a certain sense of adventure and challenge.

Final Remarks

The present chapter contributes to understanding socio-cultural impacts of sport events, a field which is gaining more and more attention in the decision-making process when bidding for events. The results of this research provide insights in behaviors and profiles of the visitors, which can be useful information for stakeholders involved in organizing cycling events.

This study demonstrates that people who are more active and participate in tours are more likely to be more motivated to attend a major cycling event. Their level of participation and engagement is higher, and this can be used in terms of building loyalty to the sport and the sport event.

Additionally, attention should focus on increasing women's interest and participation. Catering to their needs and motivations, such as gaining experience and social bonding, can be important in making such an event more meaningful and inclusive to women.

Event managers as well as local administration should also be aware that the event serves social motives, becoming a platform for interaction. When considering the event as a social occasion, it's clear that the facilitation for social interaction can be improved for these cycling tours and other events.

Overall, repeated visitors who will participate in cycling activities are committed to their health and well-being. They are then supported, and these activates are reinforced, by the social dimension, which contributes to well-being and quality of life. However, further research is needed to understand how contextual factors play a role in motivation and further participation in cycling events, and to what extent these contribute to placemaking.

References

Adrian, B., Chris, R., Heather, B., Bauman, A., Rissel, C., & Bowles, H. (2006). Mass community cycling events: Who participates and is their behaviour influenced by participation? *International Journal of Behavioral Nutrition and Physical Activity, 3*(1), 39. https://doi.org/10.1186/1479-5868-3-39

Aldred, R., & Jungnickel, K. (2014). Why culture matters for transport policy: The case of cycling in the UK. *Journal of Transport Geography, 34*, 78–87. https://doi.org/10.1016/j.jtrangeo.2013.11.004

Alexandris, K., Konstantinos, A., & Alexandris, K. (2012). Exploring the role of motivation on the development of sport involvement. *International Journal of Sport Management and Marketing, 12*(1–2), 57.

Barratt, P. (2017). Healthy competition: A qualitative study investigating persuasive technologies and the gamification of cycling. *Health & Place, 46*, 328–336. https://doi.org/10.1016/j.healthplace.2016.09.009

Berridge, G. (2012a). The promotion of cycling in London: The impact of the 2007 Tour de France Grand Depart on the image and provision of cycling in the capital. *Journal of Sport and Tourism, 17*(1), 43–61. https://doi.org/10.1080/14775085.2011.635018

Berridge, G. (2012b). Impact of 2007 Tour de France Grand Depart on cycling in London. *4th Australian conference* (pp. 42–56).

Bonham, J., & Lumb, P. (2012). Australian cycling conference 2012. In *Proceedings of the 4th Australian Cycling conference*.

Brown, T. D., O'Connor, J. P., & Barkatsas, A. N. (2009). Instrumentation and motivations for organised cycling: The development of the cyclist motivation instrument (CMI). *Journal of Sports Science & Medicine, 8*(2), 211–218.

Chatterjee, K., Sherwin, H., & Jain, J. (2013). Triggers for changes in cycling: The role of life events and modifications to the external environment. *Journal of Transport Geography, 30*, 183–193. https://doi.org/10.1016/j.jtrangeo.2013.02.007

Davies, L., Coleman, R., & Ramchandani, G. (2011). Measuring attendance: Issues and implications for estimating the impact of free-to-view sports events. *International Journal of Sports Marketing and Sponsorship, 12*(1), 11–23. https://doi.org/10.1108/ijsms-12-01-2010-b003

Deenihan, G., & Caulfield, B. (2015). Do tourists value different levels of cycling infrastructure? *Tourism Management, 46*, 92–101. https://doi.org/10.1016/j.tourman.2014.06.012

Derom, I., & Ramshaw, G. (2016). Leveraging sport heritage to promote tourism destinations: The case of the Tour of Flanders Cyclo event. *Journal of Sport and Tourism, 20*(3–4), 263–283. https://doi.org/10.1080/14775085.2016.1212393

Derom, I., & Van Wynsberghe, R. (2015). Extending the benefits of leveraging cycling events: Evidence from the Tour of Flanders. *European Sport Management Quarterly, 15*(1), 111–131. https://doi.org/10.1080/16184742.2014.997772

Derom, I., Van Wynsberghe, R., & Scheerder, J. (2015). Maintaining physical activity post-event? Case of the Tour of Flanders Cyclo in Belgium. *Annals of Leisure Research, 18*(1), 25–47. https://doi.org/10.1080/11745398.2014.932699

du Preez, E. A., & Heath, E. T. (2016). Determining the influence of the social versus physical context on environmentally responsible behaviour among cycling spectators. *Journal of Sport and Tourism, 20*(2), 123–143. https://doi.org/10.1080/14775085.2016.1227274

Fullagar, S., & Pavlidis, A. (2012). "It's all about the journey": Women and cycling events. *International Journal of Event and Festival Management, 3*(2), 149–170. https://doi.org/10.1108/17582951211229708

Funk, D., Jordan, J., Ridinger, L., & Kaplanidou, K. (2011). Capacity of mass participant sport events for the development of activity commitment and future exercise intention. *Leisure Sciences, 33*(3), 250–268. https://doi.org/10.1080/01490400.2011.564926

Garrard, J., Rose, G., & Lo, S. K. (2008). Promoting transportation cycling for women: The role of bicycle infrastructure. *Preventive Medicine, 46*(1), 55–59. https://doi.org/10.1016/j.ypmed.2007.07.010

Gibson, H., & Chang, S. (2012). Cycling in mid and later life: Involvement and benefits sought from a bicycle tour. *Journal of Leisure Research, 44*(1), 23–51. https://doi.org/10.1080/00222216.2012.11950253

Griffin, G. P., & Jiao, J. (2015). Crowdsourcing bicycle volumes : Exploring the role of volunteered geographic information and established monitoring methods. *Compendium of Transportation Research Board Annual Meeting, 27*(1), 57–66.

Halpenny, E. A. (2010). Pro-environmental behaviours and park visitors: The effect of place attachment. *Journal of Environmental Psychology, 30*(4), 409–421. https://doi.org/10.1016/j.jenvp.2010.04.006

Halpenny, E. A., Kulczycki, C., & Moghimehfar, F. (2016). Factors effecting destination and event loyalty: Examining the sustainability of a recurrent small-scale running event at Banff National Park. *Journal of Sport and Tourism, 20*(3–4). https://doi.org/10.1080/14775085.2016.1218787

Herman, Z. T. (2015). Serious leisure and leisure motivations among self-identified cyclists. *Journal of Tourism and Hospitality Management, 3*(1–2), 32–40.

Hinch, T., & Holt, N. L. (2017). Sustaining places and participatory sport tourism events. *Journal of Sustainable Tourism, 25*(8). https://doi.org/10.1080/09669582.2016.1253703

Hungenberg, E., Gray, D., Gould, J., & Stotlar, D. (2016). An examination of motives underlying active sport tourist behavior: A market segmentation approach. *Journal of Sport and Tourism, 20*(2), 81–101. https://doi.org/10.1080/14775085.2016.1189845

Jamieson, N. (2014). Sport tourism events as community builders-how social capital helps the "locals" cope. *Journal of Convention and Event Tourism, 15*(1), 57–68. https://doi.org/10.1080/15470148.2013.863719

Kaplanidou, K., & Vogt, C. (2007). The interrelationship between sport event and destination image and sport tourists' behaviours. *Journal of Sport and Tourism, 12*(3–4), 183–206. https://doi.org/10.1080/14775080701736932

King, K., Shipway, R., Lee, I. S., & Brown, G. (2018). Proximate tourists and major sport events in everyday leisure spaces. *Tourism Geographies, 20*(5), 880–898. https://doi.org/10.1080/14616688.2018.1477827

Koruç, P. B. (2015). Does exercising for a while changes the motivation of exercise participation? *International Journal of Science Culture and Sport, 3*(3), 25–31. https://doi.org/10.14486/I

Kulczycki, C., & Halpenny, E. A. (2014). Sport cycling tourists' setting preferences, appraisals and attachments. *Journal of Sport and Tourism, 19*(2), 169–197. https://doi.org/10.1080/14775085.2015.1070741

Kyle, G., Absher, J., Norman, W., Hammitt, W., & Jodice, L. (2007). A modified involvement scale. *Leisure Studies, 26*(4), 399–427. https://doi.org/10.1080/02614360600896668

Lonsdale, C., Hodge, K., Hargreaves, E. A., & Ng, J. Y. Y. (2014). Comparing sport motivation scales: A response to Pelletier et al. *Psychology of Sport & Exercise, 15*(5), 446–452. https://doi.org/10.1016/j.psychsport.2014.03.006

Lonsdale, C., Hodge, K., & Rose, E. A. (2008). The behavioral regulation in sport questionnaire (BRSQ): Instrument development and initial validity evidence. *Journal of Sport and Exercise Psychology, 30*(3), 323–355. https://doi.org/10.1123/jsep.30.3.323

Mackellar, J., & Jamieson, N. (2015). Assessing the contribution of a major cycle race to host communities in South Australia. *Leisure Studies, 34*(5), 547–565. https://doi.org/10.1080/02614367.2014.938772

McIntyre, N. (1989). The personal meaning of participation: Enduring involvement. *Journal of Leisure Research, 21*(2), 167.

Meschik, M. (2012). Sustainable cycle tourism along the Danube Cycle route in Austria. *Tourism Planning and Development, 9*(1), 41–56. https://doi.org/10.1080/21568316.2012.653478

Moon, K. S., Ko, Y. J., Connaughton, D. P., & Lee, J. H. (2013). A mediating role of destination image in the relationship between event quality, perceived value, and behavioral intention. *Journal of Sport and Tourism, 18*(1), 49–66. https://doi.org/10.1080/14775085.2013.799960

Navarro, K. F., Gay, V., Golliard, L., Johnston, B., Leijdekkers, P., Vaughan, E., Wang, X., & Williams, M. (2013). SocialCycle what can a mobile app do to encourage cycling? *38th Annual IEEE conference on local computer networks – workshops* (pp. 24–30). https://doi.org/10.1109/LCNW.2013.6758494

O'Connor, J. P., & Brown, T. D. (2007). Real cyclists don't race: Informal affiliations of the weekend warrior. *International Review for the Sociology of Sport, 42*(1), 83–97. https://doi.org/10.1177/1012690207081831

Pelletier, L. G., Rocchi, M. A., Vallerand, R. J., Deci, E. L., & Ryan, R. M. (2013). Validation of the revised sport motivation scale (SMS-II). *Psychology of Sport and Exercise, 14*(3), 329–341. https://doi.org/10.1016/j.psychsport.2012.12.002

Pelletier, L. G., Tuson, K. M., Fortier, M. S., Vallerand, R. J., Briére, N. M., & Blais, M. R. (1995). Toward a new measure of intrinsic motivation, extrinsic motivation, and amotivation in sports: The sport motivation scale (SMS). *Journal of Sport and Exercise Psychology, 17*(1), 35–53. https://doi.org/10.1123/jsep.17.1.35

Reitsamer, B. F., Brunner-Sperdin, A., & Stokburger-Sauer, N. E. (2016). Destination attractiveness and destination attachment: The mediating role of tourists' attitude. *Tourism Management Perspectives, 19*, 93–101. https://doi.org/10.1016/j.tmp.2016.05.003

Rejón-Guardia, F., García-Sastre, M. A., & Alemany-Hormaeche, M. (2017). Motivation-based behaviour and latent class segmentation of cycling tourists: A study of the Balearic Islands. *Tourism Economics, 24*(2), 204–217. https://doi.org/10.1177/1354816617749349

Ritchie, B. W., Shipway, R., & Cleeve, B. (2009). Resident perceptions of mega-sporting events: A non-host city perspective of the 2012 London olympic games. *Journal of Sport and Tourism, 14*(2–3), 143–167. https://doi.org/10.1080/14775080902965108

Ritchie, B. W., Tkaczynski, A., & Faulks, P. (2010). Understanding the motivation and travel behavior of cycle tourists using involvement profiles. *Journal of Travel and Tourism Marketing, 27*(4), 409–425. https://doi.org/10.1080/10548408.2010.481582

Rocha, C. M., & Gratao, O. A. (2018). The process toward commitment to running—The role of different motives, involvement, and coaching. *Sport Management Review, 21*(4), 459–472. https://doi.org/10.1016/j.smr.2017.10.003

Rose, G., & Marfurt, H. (2007). Travel behaviour change impacts of a major ride to workday event. *Transportation Research Part A: Policy and Practice, 41*(4), 351–364. https://doi.org/10.1016/j.tra.2006.10.001

Rowe, K., Shilbury, D., Ferkins, L., & Hinckson, E. (2016). Challenges for sport development: Women's entry level cycling participation. *Sport Management Review, 19*, 417–430. http://doi.org/10.0.3.248/j.smr.2015.11.001

Ryan, R., Frederick, C., Lepes, D., Rubio, N., & Sheldon, K. (1997). Intrinsic motivation and exercise adherence. *International Journal of Sport Psychology, 28*(4), 335–354.

Ryan, R. M., & Deci, E. L. (1985). *Self-determination theory: Basic psychological needs in motivation, development, and wellness.* The Guilford Press.

Schulenkorf, N. (2009). An ex-ante framework for the strategic study of social utility of sport events. *Tourism and Hospitality Research, 9*(2), 120–131. https://doi.org/10.1057/thr.2009.2

Smith, A. C. T. T., & Stewart, B. (2007). The travelling fan: Understanding the mechanisms of sport fan consumption in a sport tourism setting. *Journal of Sport and Tourism, 12*(3–4), 155–181. https://doi.org/10.1080/14775080701736924

Smith, W. R., & Treem, J. (2017). Striving to be king of mobile mountains: Communication and organizing through digital fitness technology. *Communication Studies, 68*(2), 135–151. http://doi.org/10.0.4.56/10510974.2016.1269818

Spinney, J. (2006). A place of sense: A kinaesthetic ethnography of cyclists on Mont Ventoux. *Environment and Planning D: Society and Space, 24*(5), 709–732. https://doi.org/10.1068/d66j

Spotswood, F., Chatterton, T., Tapp, A., & Williams, D. (2015). Analysing cycling as a social practice: An empirical grounding for behaviour change. *Transportation Research Part F: Traffic Psychology and Behaviour, 29*, 22–33. https://doi.org/10.1016/j.trf.2014.12.001

Stragier, J., Evens, T., & Mechant, P. (2015). Broadcast yourself: An exploratory study of sharing physical activity on social networking sites. In *Media International Australia* (Vol. 155, pp. 120–129).

Sun, Y. (2017). Exploring potential of crowdsourced geographic information in studies of active travel and health: STRAVA data and cycling behaviour. *The International Archives of the Photogrammetry, Remote Sensing and Spatial Information Sciences, XLII-2-W7*, 1357–1361. https://doi.org/10.5194/isprs-archives-XLII-2-W7-1357-2017

Sun, Y., Du, Y., Wang, Y., & Zhuang, L. (2017). Examining associations of environmental characteristics with recreational cycling behaviour by street-level Strava data. *International Journal of Environmental Research and Public Health, 14*(6), 644. https://doi.org/10.3390/ijerph14060644

Van Cauwenberg, J., Clarys, P., De Bourdeaudhuij, I., Van Holle, V., Verte, D., De Witte, N., ... Deforche, B. (2012). Relationships between the physical environment and older adults' walking and cycling behaviours: The Belgian aging studies. *Journal of Aging and Physical Activity, 20*, 314–315.

Ward, T. (2014). Too much information ? *Cyclist, 22*, 136.

Wardman, M., Tight, M., & Page, M. (2007). Factors influencing the propensity to cycle to work. *Transportation Research Part A: Policy and Practice, 41*(4), 339–350. https://doi.org/10.1016/j.tra.2006.09.011

West, L. R. (2015). Strava: Challenge yourself to greater heights in physical activity/cycling and running. *British Journal of Sports Medicine, 49*(15), 1024. http://dx.doi.org/10.1136/bjsports-2015-094899

Zander, A., Passmore, E., Mason, C., & Rissel, C. (2013). Joy, exercise, enjoyment, getting out: A qualitative study of older people's experience of cycling in Sydney, Australia. *Journal of Environmental & Public Health, 2013*, 1–6.

11 Mountaineering Management

Jaroslav Kupr and Klara Kuprova

Introduction

The aim of this chapter is to bring attention to the specific conditions of mountain stays and to explain the importance of mountaineering management in realization of projects placed in a mountain environment. For better understanding, we use a sample study of a mountain expedition at 7,000 m in elevation.

Mountain environments, and the activities placed in them, have many specific conditions. The specific differences in environment (e.g., altitude and distance of lodging) put high requirements on all athletes, tourists, and others. It is necessary to take into account all kinds of risks, which individuals do not encounter in everyday life.

Mountaineering is a human activity with rich history and tradition, which constantly pushes the limits of human possibilities and performances. Mountaineering could be compared with other professional sports, but often it does not get much attention from media and society. However, it is open to a variety of people, who come to mountains for personal experience and pleasure. This chapter will examine the management of risk, subjective and objective danger, organizational factors of mountaineering projects, assessment of risks in a mountain environment and more. The main goal of this chapter is to highlight aspects of the preparation and execution of expeditions with emphasis on minimizing risk for participants and considering their individual needs.

Definition of Mountaineering

The broad term *mountaineering* includes many insights and opinions on the sport. In its original meaning, mountaineering represents a climbing activity with a goal to reach the summit of the mountain (Horolezecká metodika, 2010; Procházka, 1975). However, today the term *climbing* contains many activities and their specific disciplines. Vomáčko and Boštíková (2003) divide traditional climbing activities into rock climbing, sandrock climbing, bouldering, big walls, ice-mix climbing, and high mountaineering. We can also include some outdoor activities related to climbing, such as hiking, skiing, via ferratas, and many more. The mutual component of all these activities is the movement in a mountain terrain.

DOI: 10.4324/9781003476658-14

A separate branch of climbing evolution is sport climbing. Sport climbing is a competitive discipline of climbing (e.g., lead climbing, speed climbing, bouldering, and ice climbing). A significant step for the evolution of climbing was the Olympic Games in Tokyo 2020 (realized in 2021) (ČHS, 2021). The climbing was for the first time as a part of the Olympic Games. It was composed of three disciplines (boulder, lead, and speed), and the final rankings were determined by computing the product of each climber's rankings in the three disciplines (Saito et al., 2022; Stinson & Stinson, 2021). This popularization was a significant and positive impact on sports climbing in the present. However, sport climbing involves only a small group of athletes focused on performance. More important is the effect of mountaineering on crowds of non-competitive individuals. It is always interesting to see what non-competitive mountaineers have in their motivation, enjoyment, personal experience, etc.

History of Mountaineering

Big mountains have always been a part of human life. Holy mountains, which many see as being inhabited by Gods, can be found on practically all continents (e.g., Mt. Olympus at 2,917 m, Mt. Nanda Devi at 7,816 m, and Mt. Kailash at 6,713 m). Therefore, the fascination with mountain peaks could have been as old as mankind itself (Dieška & Širl, 1989).

The beginning of climbing history can be dated to the end of the 18th century. A major breakthrough was made by conquering the summit of Mont Blanc (4,807 m) in 1786, followed by other peaks – Grossglockner (3,797 m) in 1800, and Ortler (3,899 m) in 1804. In the mid-19th century, the English began to explore more difficult and not-yet-conquered summits. At the same time, the first climbing association, the British Alpine Club, was created in London, 1857. Since 1863, the association regularly published a magazine called *Alpine Journal* (Dieška & Širl, 1989). A significant date is 14 July 1865, when the Mount Matterhorn was conquered for the first time (Messner, 2017). Since that time, mountaineers have been constantly pushing the limits of climbing possibilities and difficulties. Furthermore, it gave rise to the interest in the highest mountain peaks of the world.

In 1848, people systematically measured the mountains' elevation in the Himalayas. In 1852, they determined the highest peak of the world, named Peak XV. Consequently, it was then renamed to Chomolungma, Sagarmatha, and finally to Mount Everest. The first attempt to conquer the summit was dated to 1921. Likely the most famous name related to Mount Everest is G. L. Mallory. Mallory and A. Irvin died during one of their efforts to reach the peak in 1924. The first successfully conquered 8,000 m peak was Annapurna in 1950, followed by Mount Everest in 1953, Nanga Parbat in 1953, and many more, the last conquered 8,000 m peak being Shisha Pangma in 1964 (Jasanský & Rakoncaj, 2003; Rakoncaj, 2003).

After all these successful 8,000 m peak expeditions, people started to search for new routes to the top of these mountains. One of the most significant people who pushed the human limits in climbing was undoubtedly Reinhold Messner (born 1944). He is the first person to conquer the summit of Mount Everest without supplemental oxygen, alongside Peter Habeler in 1978. Messner performed a first individual climb on Mount Everest in 1980 and completed all fourteen 8,000 m peaks in 1980. In the same year he completed the project called 7 Summits (4th man overall and 1st without supplemental oxygen) (Baláš, 2016; Messner, 2006).

At the end of the 20th century and beginning of the 21st century, we can see new trends and approaches that were previously unfeasible. Now, it is preferred to use simpler styles of expeditions with a small number of participants and minimal resources or "Alpine style" without any support or building sequential camps. Furthermore, many climbers now search new possible climbing routes or others use skis or even paraglide to descent. On the other hand, some mountains became very commercially attractive (e.g., Mt. Everest) and the style of summitting is now quite different than the original expeditions. However, it can be stated that climbing in mountains, at all levels, is currently a very attractive leisure activity to many people around the world. This is demonstrated through the increasing number of members in climbing associations within numerous countries (e.g., Czech Mountaineering Association with 20,000 members and Alpenverein with 500,000 members) (Alpenverein, 2018).

Motivation for Mountaineering

The desire to explore and overcome new challenges is as old as mankind itself. Mountains have always attracted humans and the aspiration to discover what is hidden in their core. Gradually, humans have managed to conquer the Alpine peaks, difficult summits in Pamir at Kavkaz, and finally all the 8,000 m peaks. Nevertheless, exploration is not at its end. It is still possible to discover new climbing routes and approaches, which were previously impossible. Mainly, it is possible to explore the inner human motivation which attracts people to the mountains.

Not everyone is able to reach the peak of the highest mountain in the world or find a new and more difficult climbing route to push human limits. However, non-competitive mountaineers share the same impulses and motivation as those who belong to top elite climbers. G. L. Mallory once answered the question, "Why do you climb Mt. Everest?" quite simply: "Because it is there" (Messner, 1999).

Dina Štěrbová (1988), the first Czech woman to conquer an 8,000 m peak, says that the happiness from achieving goals is equal to the sacrifice on the way to it. Czech climber Josef Nežerka says that mountains make him a better person. Maybe it is these thoughts and many more that push a person to make expeditions to environments which are not always hospitable and can even prove dangerous. Personal sacrifices in the forms of exertion, time, energy, money, and more should be admired. Despite the sacrifice, climbing a mountain gives humans a feeling full of satisfaction and inner peace (Baláš, 2016; Holeček, 2018; Jaroš, 2018) (Figure 11.1).

Figure 11.1 Satisfaction after Reaching the Peak Lenin in 2013 (7,134 m).

Management

Management is a process of systematic planning, organizing, decision-making, controlling, as well as communication and leadership. All roles have one common intention: to set and achieve a goal by using all available resources (e.g., financial, natural, technological, and human resources). Management also includes activities, such as setting up a strategy for organization and coordination of employees or volunteers. The term *management* can also refer to people who lead the organization (Veber, 2014).

Risk Management

Risk management is a system of projects and processes that are concerned with the recognition and evaluation of an activity's risks and undesirable consequences. It is a systematic, repetitive set of mutually connected activities, whose goal is to limit the probability of risk occurrence or reduce its consequences on the project (Loská & Kabálková, 2006). The evaluation of risk is defined as a complex process of determining the magnitude and probability of creating an undesirable situation and decision, where arrangements must be made to eliminate or reduce risk to an acceptable degree (Bernatík, 2016; Pokorná, 2008). The purpose of risk management is to prevent undesirable surprises and to keep problems from escalating. In outdoor projects such as high-altitude mountain expeditions, risk management processes usually focus on the identification and elimination of subjective and objective risks.

Subjective Risk

A subjective risk is the sum of impacts dependent on the subject's activity. In this context we talk about a climber being the subject. The model example could be, for example, an attempt to climb a 7,000 m peak.

Percentage-wise, most climbing injuries are caused by a subjective risk (ČHS, 2018; Kublák, 2021), which are usually one's own mistakes and failings. Through prevention (e.g., training and preparation) it is possible to reduce or even eliminate a potential risk. Subjective risk could be divided into six categories:

1 Physical and technical preparation

Sufficient performance during an expedition is dependent on proper and specified preparation. A big emphasis is put on one's endurance, technical (climbing) training, and on specific project requirements (e.g., ice-climbing and ice mix-climbing). It is very important for mountaineers to experience high altitude, cold, and other extreme factors (Dovalil, 2009; Perič & Dovalil, 2010). Most of the time, individuals overestimate their own strength and abilities in these categories. Since one's reaction to altitude is so specific and individual, it is appropriate to test one's reaction using a long-term process before the expedition.

2 Psychological training

The psychological training for an expedition is just as important as the physical training. In some specific phases of an expedition, an individual's state of mind can decide between success and failure. It is necessary to practice physical and emotional exertion under control to examine situations correctly and avoid rushed and fatal decisions. There is evidence that proves altitude has a negative effect on human's mental health due to insufficient amount of oxygen and local hypoperfusion of the brain. In extreme situations, it can cause hallucinations, confusion, concentration disturbances, etc.

3 Specific skills training

The planned mountaineering project can require that participants have specific skill training before attending. New skills should be trained, effective, and automatic. This training is sometimes underestimated or insufficient. The skills training can contain activities, such as walk in crampons, belaying, rescue from a crack, orientation in a terrain, etc. Insufficient skills training is commonly seen by commercial expeditions, which combine heterogeneous groups of climbers with different levels of experience.

4 Material preparation

Material preparation is one of the necessary parts of project preparation. All climbing material must be tested, regulated, and adjusted to individual needs. Failure or false manipulation of material can have fatal consequences.

5 Group training

Group/team training is very necessary during expeditions that require cooperation or individual specialization of some skills. The whole project could be affected by many group characteristics, such as practicality of actions, forms of communication, or sharing of information.

6 Financial preparation

Many projects in high mountains are financially demanding. Therefore, it is necessary to calculate the required amount of money in advance to prevent personal loss (including personal resources as well as sponsor donations). The financial requirements of the project can include resources from before and during the expedition as well as some lost profit. In many destinations around the world, it is required to have enough cash for a sudden risk situation (rescue, transportation, help from locals, etc.).

7 Human factors

A human factor or failure can cause harm to the individual as well as the rest of the group (injury, loss of material, altitude sickness, etc.).

Objective Risk

Objective risk is a sum of impacts independent of the subject's activity. These negative consequences can be eliminated by early examination of the specific situation, pre-training, individual experience, training of crisis situations, etc. The objective risks in climbing environments are the circumstances surrounding the climber, not the climber themselves. A good example of objective risk is an effort to reach a peak of 7,000 m in altitude.

Objective risks could be separated into five categories:

1 Location risks

Every location has its own specific risks, such as war or political stability, probability of terrorist attacks, language barriers, differences in religion, and more. All these things can lead to a very dangerous situations which can be hard to solve on site.

2 Terrain risks

Terrain and environment can suddenly change a project's character and turn it into a dangerous situation. As terrain risks, we consider earthquake, tornado, avalanche (snow or rock), fallen trees, melting glaciers, etc.

3 Weather changes

Sudden and unplanned changes in weather can have negative or even fatal effect on the whole expedition. Even during the best examination of a specific trek, the effect of weather is important and often unpredictable (e.g., change of temperature, snow conditions, visibility).

4 Material loss

Even when participants have sufficient climbing material, the material can get lost, damaged, or destroyed. Some materials require numerous spares or a universal repair kit to prevent crisis situations. Some losses (e.g., tent, sleeping bag, special shoes) are not possible to replace with alternatives and are crucial for a successful expedition and to preserve one's life.

Mountaineering Management

The planning of a project by itself is a process which begins well in advance, before the expedition even starts. There are many phases of decision-making and planning which can postpone the project or even reject it (Nová et al., 2016). Reasons for rejection could be due to both objective and subjective risks to a climber. In general, we can define four stages of the planning process: a view from home (planning), a view from a window (realization of the project), a view from the spot (realization of the expedition peak), and evaluation (Horolezecká metodika, 2010; Kublák, 2021).

1   Stage 1 – A View from Home

   During the first stage of planning, it is important to ask what the original impulse for this project is. Motivation for the realization of a project could be internal (e.g., a longtime dream/plan, personal development, self-realization, personal pleasure) or external (e.g., an impulse from individual's background, pressure from sponsors, an invitation from colleagues).

   The final decision to realize a project is affected by information analysis. The information includes examination of the climb difficulty, financial demands, time of year, logistic demands, time management, vaccination, transportation and cargo express, materials, communication services, political situation, seasonal weather, and snow conditions. This information is collected from many different sources (e.g., internet, discussion forum, friends, maps, technical magazines, photos) in an effort to maximize data and make a better final decision. An example of an expedition budget is shown in Table 11.1 and an example of material requirements in Table 11.2.

   After a detailed information analysis, a final decision is made – YES or NO. Many projects perish or are postponed in this early phase for many different reasons.

   If the decision was positive, a pre-departure preparation is needed to eliminate all subjective risks (see list of subjective risks above) for a climber. It is necessary to have enough time to go through all subjective factors before departure.

2   Stage 2 – A View from a Window

   The second stage of the planning process involves realizing the project itself. In this phase, subjects are already at the starting point of their expedition and must evaluate its situation. The evaluation of the current situation is affected by many factors (e.g., weather forecast, physical and mental health of subjects, group dynamics, condition of the materials, time management, snow conditions), followed by a decision of whether or not the expedition should commence. Evaluation of both objective and subjective risks is crucial for the whole project. At this point, it is good to have an alternative plan if leaders decide not to start the expedition.

3   Stage 3 – A View from the Spot

   The third stage is no longer about planning, but about examining the real situation in the place of expedition (when climbing already has begun). The key factor in this stage is to evaluate all accessible information whenever possible

Table 11.1 Timeline

Activity	Date
First thought about an expedition	September 2014
First collection of information, finances, and time balance	October 2014
Information from friends, more in-depth analysis	November 2014
Final decision to make an expedition	December 2014
Addressing an agency for mediation of services (Kyrgyzstan)	January 2015
Purchase of flight tickets	February 2015
Deposit payment to the agency (Kyrgyzstan)	February 2015
Ski training in the Alps (Austria), group training	March 2015
Collection of materials	April 2015
Submit an application for visa to China	May 2015
Tests and control of materials	June 2015
Take out insurance contracts	June 2015
Physical and mental training	October 2014–July 2015
Obtaining a visa to China	July 2015
Departure to expedition	End of July 2015
The expedition	28 days until expedition
Return home	End of August 2015
Evaluation	September 2015
Presentation of the expedition (pictures and video) for friends and partners of the project	October 2015

Table 11.2 Expedition Expenses

Item	Total price (Czech crowns)	Total price (EUR)
Flight ticket	13,000	500
Permit, agency	67,500	2,596
Other payments	3,500	135
Insurance	6,970	268
Vaccination	2,200	85
Visa to China	1,500	58
Total expenditure	94,670	3,641

(analyzing weather, participant health, material shape, route directions, etc.). After a detailed analysis of accesses to the climbing area, it is possible to realize the whole (prepared) plan or pass to a modified or alternative plan.

4 Stage 4 – Evaluation

Evaluation is an inseparable part of all projects (Hájek et al., 2008). There are two possible approaches to evaluation – external evaluation and internal (self) evaluation. External evaluation rates the success of a project from an external perspective (e.g., conquering the peak, public relations, and successful presentation). Internal evaluation comes from an individual's feelings such as their satisfaction with the whole expedition and personal experiences. However, external and internal evaluations do not have to come to the same conclusion (Hendl & Reml, 2017).

Timeline of an Expedition

Table 11.1 displays a detailed timeline for the China expedition in 2015. The goal of this expedition was to summit Mount Muztagh Ata (7,546 m) in the Pamir Mountain Range (43rd highest mountain in the world). Both ascent and descent were successfully completed on skis. The expedition involved only two men, both of them were experienced climbers from previous expeditions, but this mountain was at their high-altitude maximum. The whole process of preparation took them around 11 months, prior to the actual start of the expedition.

Example of Financial Demands

In Table 11.2, you can see a detailed calculation of expenses for the China expedition in 2015. In the table are direct expenses used during the expedition itself. A permit to the mountain and a service provision were provided by the agency from Kyrgyzstan through a liaison officer in China. They did not include expenses prior to the beginning of the expedition.

Example of Material Demands

Specific equipment for this expedition were skis for ski mountaineering (both climbing and downhill) (Table 11.3). Other equipment is similar to materials used during a normal climbing expedition (for 7,000 and 8,000 m peaks). Necessary and expensive items are down clothes (jacket and gloves), sleeping bags, and a special tent for high altitudes. Climbers also need high-quality goggles and electronics such as a headlamp, camera, and watch. It is fundamental to have a satellite phone which allows communication independent of mobile network. Climbing equipment is usually set up for winter conditions. We place a big emphasis on wearing material that is made to be as light as possible to ensure everything is functional and safe. The total weight of material used for this expedition was 37 kg.

Extreme Mountaineering

An extreme approach to climbing can be understood through different perspectives. The first perspective is an effort to push human limits and the second is about focusing on the performance itself (Hoffmanová & Šebek, 2013). The effort to push human limits is closely related to climbing and conquering mountain summits. It could already be considered a revolutionary approach to staying in the mountains. Currently, climbing performances and records are now overcome more by speed than difficulty. Mountaineers, such as Uli Steck (speed record in a climb on Eiger), Killian Jornet (two successful climbs to Mt. Everest in one week), Marek Holeček (new climbing route on Gasherbrun I) and many more, push the human limits to unbelievable levels.

Therefore, the word *extreme* gains a whole new meaning, for some to draw attention, to astound, or to promote something. The motivation behind these performances is heterogeneous, but it does not decrease the attractiveness of similar performances.

Table 11.3 Material Demands for the China Expedition in 2015

Item	Total price (Czech crowns)	Total price (EUR)
Sleeping		
Tent, sleeping pad, sleeping bag, sleeping boots	19,300	742
Cooking		
Jet-Boil, spoon, lighter, mug, knife, thermos, water bottle	4,200	162
Skiing equipment		
Boots, poles, irons, 2× skiing belts, avalanche transceiver, helmet, shovel, goggles, skis	47,500	1,827
Backpacks	9,900	381
Electronics		
Solar panel, battery, cell phone, MP3, Kindle, headlamp, watch, camera, SD card, cables, satellite phone	57,000	2,192
Glasses	3,200	123
Boots		
Trekking boots, sandals, sleeves	6,800	262
Underwear		
5× socks, 4× shorts, 2× long underpants	5,000	192
T-shirts	4,900	188
Hoodies		
2× sweatshirt, vest, down vest	7,510	289
Jackets		
Rain jackets, down Jacket	16,900	650
Pants		
Shorts, rain pants, ski pants	18,500	712
Gloves		
Thin gloves, down mittens, ski gloves	7,500	288
Hats and bandanas	2,950	133
Hygiene	700	27
Meals for altitude climbing	4,200	162
Service material	265	10
Climbing material		
Harness, 3× carbines, T-block, rope, crampons, ice axes, poles	7,700	296
First aid kit	4,500	173
Total expenditure	228,525	8,809

Conclusion

The aim of this chapter is to bring attention to specific conditions of mountain trekking and explain the importance of mountaineering management in realization of projects in a mountain environment. For better understanding, we used a sample study of a 7,000 m peak expedition. This chapter describes the necessity of being aware of possible risks in a mountain environment, with an emphasis on the terms *risk management, subjective* and *objective risk,* and *their division to individual groups.* This chapter also mentions four organizational phases of a project from a strategic view. A large part is devoted to specific conditions and demands, which are caused by the high-altitude mountain environments. Finally, mountaineering is a human activity with rich history and tradition, which constantly pushes the limits of human possibilities and performances. However, it is open to various people who come to the mountains for personal experience and pleasure.

References

Alpenverein. (2018). *Historie [History]*. Retrieved from https://www.oeav.cz/historie

Baláš, J. (2016). *Fyziologické aspekty výkonu ve sportovním lezení [Physiological aspects of performance in sport climbing]*. Praha: Univerzita Karlova.

Bernatík, A. (2016). *Analýza nebezpečí a rizik [Hazard and risk analysis]*. Retrieved from https://www.fbi.vsb.cz/export/sites/fbi/U3V/cs/materialy/U3V_AnalyzaRizik.pdf.

ČHS. (2018). *Historie horolezectví [Climbing history]*. Retrieved from Český horolezecký svaz website: https://www.horosvaz.cz/historie-horolezectvi/

ČHS. (2021). *Olympic sport climbing*. Retrieved from Český horolezecký svazecký svaz website: https://www.horosvaz.cz/olympijske-hry/

Dieška, I., & Širl, V. (1989). *Horolezectví zblízka [Mountaineering close up]*. Praha: Olympia.

Dovalil, J. (2009). *Výkon a sportovní trénink [Performance and sports training]*. Praha: Olympia.

Hájek, B., Hofbauer, B., & Pávková, J. (2008). *Pedagogické ovlivňování volného času [Pedagogical influencing of free time]*. Praha: Portál.

Hendl, J., & Reml, J. (2017). *Metody výzkumu a evaluace [Research and evaluation methods]*. Praha: Portál.

Hoffmanová, J., & Šebek, L. (2013). *Fenomén X-sportů a aktivit [The phenomenon of X-sports and activities]*. Olomouc: Univerzita Palackého v Olomouci.

Holeček, M. (2018). *Prvovýstupy [First ascents]*. Retrieved from marekholecek.cz website: http://marekholecek.cz/maara.htm#subpage_26

Horolezecká metodika. (2010). *Historie horolezectví [Climbing history]*. Retrieved from Učebnice website: http://horolezeckametodika.cz/ucebnice/horolezectvi-a-sport/historie-horolezectvi

Jaroš, R. (2018). *Koruna Himáláje [Crown of the Himalayas]*. Retrieved from radekjaros.cz website: https://www.radekjaros.cz/koruna-himalaje

Jasanský, M., & Rakoncaj, J. (2003). *České himalájské dobrodružství [Czech Himalayan adventure]*. Praha: Knižní klub.

Kublák, T. (2021). *Plánování túry [Tour planning]*. Retrieved from Horolezecká metodika website: https://horolezeckametodika.cz/planovani-tury

Loská, Š., & Kabálková, P. (2006). *Risk management [Risk management]*. Retrieved from http://ikaros.cz/node/12281

Messner, R. (1999). *Mallorys zweiter Tod. Das Everest-Rätsel und die Antwort [Mallory's second death. The Everest mystery and the answer]*. München: BLV.

Messner, R. (2006). *Mein Weg [My way]*. München: Frederking & Thaler.

Messner, R. (2017). *Absturz des Himmels [Crash of the sky]*. Frankfurt a. M.: S. Fisher.

Nová, J., Novotný, J., Racek, O., Rektořík, J., Sekot, A., Strachová, M., & Válková, H. (2016). *Management, marketing a ekonomika sportu [Management, marketing and economics of sport]*. Brno: Masarykova Univerzita.

Perič, T., & Dovalil, J. (2010). *Sportovní trénink [Sport training]*. Praha: Grada.

Pokorná, G. (2008). *Projekty – jejich tvorba a řízení*. Retrieved from ESFmoduly website: http://esfmoduly.upol.cz/publikace/projekty.pdf

Procházka, V. (1975). *Základy horolezectví* [*Basics of mountaineering*]. Praha: Olympia.

Rakoncaj, J. (2003). *Polibek nebo zatracení* [*Kiss or damn*]. Praha: Knižní klub.

Saito, M., Ginszt, M., Semenova, E., Massidda, M., Huminska-Lisowska, K., Michałowska-Sawczyn, M., ... Kikuchi, N. (2022). Genetic profile of sports climbing athletes from three different ethnicities. *Biology of Sport* (November 2021). https://doi.org/10.5114/biolsport.2022.109958

Štěrbová, D. (1988). *Čo Oju, tyrkysová hora* [*Cho Oyu, turquoise mountain*]. Praha: Olympia.

Stinson, M. J., & Stinson, D. R. (2021). *An analysis and critique of the scoring method used for sport climbing at the 2020 Tokyo Olympics* (pp. 1–16). Retrieved from http://arxiv.org/abs/2108.12635

Veber, J. (2014). *Management – Základy, moderní manažerské přístupy, výkonnost a prosperita* [*Management – Fundamentals, modern management approaches, performance and prosperity*]. Praha: Management Press.

Vomáčko, S., & Boštíková, S. (2003). *Lezení na umělých stěnách* [*Climbing on artificial walls*]. Praha: Grada Publishing, a.s.

12 Sports for Tourism

Developing Windsurfing in Taiwan

Chiung-Tzu Lucetta Tsai

Introduction

Sport tourism is an emerging segment of the world's largest and fastest-growing industry, sport tourism. It capitalizes on the relationship between tourism and the high-profile, multi-billion-dollar sports industry (Castagna et al., 2008). As a growing segment of the tourism industry, the impact of sport tourism can be categorized in four folds, namely sport and tourism as economic activities, sports activity holidays, major sport facilities and events as an attraction for visitors, and sport and tourism as part of regeneration strategies for cities and regions (Hori et al., 2008). Taiwan held the 2009 World Game, the 2010 Deaflympics, and the 2017 World University Games. As a result, the interplay between sport tourism and the economy, sociocultural dynamics, people's health, the environment, and public policy are deemed important (Khadarooa & Seetanah, 2008). Marine sport tourism is deemed important in the development of sport tourism and, therefore, developing windsurfing in Taiwan has been recognized as important to increase leisure and recreation opportunity for tourists (Chang, 2009; Chang et al., 2010). Taiwan is surrounded by the sea and there are many places suitable for this Olympic sport.

The Taiwanese government has been promoting domestic water sports in the last ten years and the island has a suitable climate and favorable conditions all year round for water sport. How to assist Taiwan's citizens to understand and become familiar with the skills, culture, and regulations related to windsurfing is a major issue, while how to strengthen windsurfing in Asia is also a challenge. This study may serve as a starting point in the exploration of how windsurfing development is formulated and delivered in a given context as well as in what form. What the author has sought to achieve in this study is to identify and explain the debates around the strategic aims of development concerned with Taiwan's windsurfing in terms of sport tourism (Tsai, 2011).

History of Windsurfing

Windsurfing refers to a water sports that uses wind and wave power to drive a rudderless and cockpit-free boat (Figure 12.1). It is environmentally friendly and conforms to international trends (Vogiatzis & De Vito, 2015). The equipment consists of a board with a dagger board, the mast, universal joint, sail, and boom

DOI: 10.4324/9781003476658-15

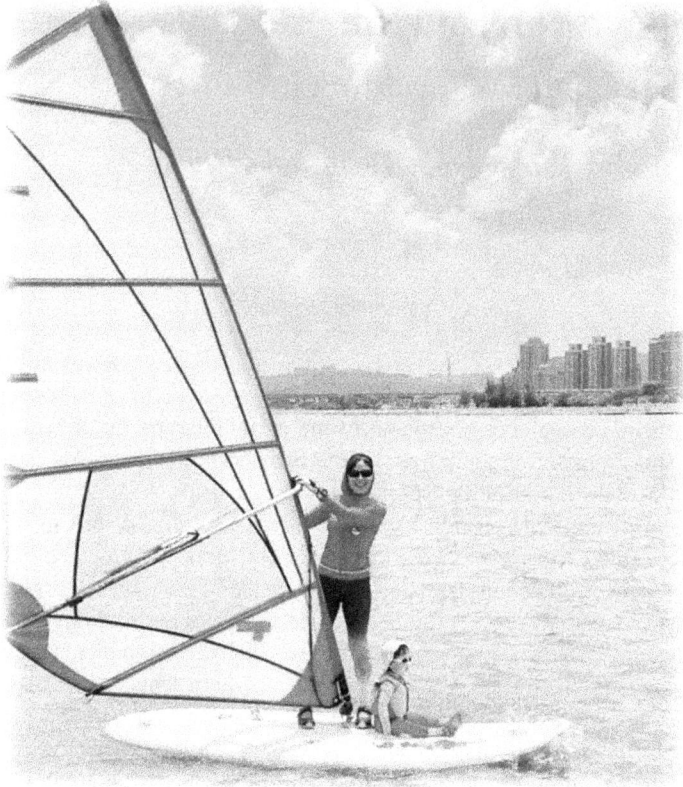

Figure 12.1 Windsurfing Is a Sport that Is Fun to Learn.
Source: Lucetta Tsai.

composition. Windsurfers use the natural wind blowing on the sail while standing on the board and operating the lever operation. Wind speeds the board on the water and movement is also generated, by changing the position of the center of gravity of the wind and sail, allowing the board to turn on the sea, river, lake, or other body of water.

Windsurfing takes the driving force of the wind and waves, combines the skills of surfing and water skiing. It can help to improve physical strength and also requires mental activity. The Olympic windsurfing competition currently uses only a triangle sailing route as the standard (Ghiani et al., 2018). Windsurfing is a level of competition in sailing, an emerging sport, is something that can easily be done independently, and the equipment is simple and light. The Mistral One Design Class (MOD) was a windsurfing class chosen by the International Sailing Federation (ISAF) for use at the Olympic Regatta in Savannah 1996, Sydney 2000, and Athens

2004. Since the 2008 Summer Olympics, the RS: X class is used for the Olympic competitions. The competition is suitable for the sea and can also be carried out in inland rivers and lakes. It originated in the Hawaiian Islands, the world's surfing paradise in the late 1960s. In 1965, Newman, an American, wrote the article "Slide Skateboarding – An Emerging Stimulating High-Speed Water Sports" for a popular science magazine, thereby introducing the installation of sails on surfboards. Windsurfing requires the participant to face the sail to operate it. As this kind of operation is not in accordance with the natural position of the human body, it is difficult to maneuver. Therefore, it did not initially attract much attention.

The first World Windsurfing Championships was held in 1974 (Nathanson, 2019; Nikolić et al., 2019). World's International Windsurfing Board Association holds several international competitions every year. In 1981, windsurfing was accepted as a member of the Olympic family as a sailing event. Since then, it has been listed as an official competition (Hori et al., 2008). Nowadays, all large-scale comprehensive sports events such as the Olympic Games, Asian Games, and National Games have windsurfing competitions (Jouke et al., 2013). Every year, regular professional series are held all over the world. Windsurfing competitions not only attract audiences, but also attract a lot of media attention and interest from businesses. Windsurfing sails can be used as an effective way to provide advertising. When people enjoy the exciting and beautiful images of windsurfing competitions on television or in newspapers, they also see the corporate logos of sponsors (Nazim et al., 2014).

Windsurfing Theory

Intensity of the Wind

Windsurfing is greatly dependent on the strength of the wind, the equipment used, and the conditions of the sea. It is not realistic to use an anemometer at sea. It is necessary for the windsurfer to use observation of various phenomena to roughly evaluate the wind speed. For beginners, noticing a wind speeds of 4–5 knots is important. If the speed is greater than five knots, then the intensity of the wind will be too strong, as a result, it will not be suitable for the operation of windsurfing activities (Verlinden et al., 2013). Windsurfers should avoid strong wind and waves at sea in the very beginning.

Understanding the Direction of Wind

The direction of the wind is also an important factor while making a windsurfing the maneuver. Even if the angle of the wind is small, as soon as the wind begins to turn from the land to the sea, the windsurfer should turn back to the shore. Windsurfing requires a certain number of operational skills when the wind blows 'downwind' or 'side downwind' (Zalaudek et al., 2019). Windsurfers should avoid participating in the sport at sea prior to fully mastering it. Changes of wind direction can also lead to changes in ocean conditions. As the wind speeds increase, the surface of the sea will become choppy and, therefore, dangerous for windsurfing.

Northeast Monsoon in Taiwan

As a kind of natural phenomenon, wind changes regularly with the changes of the four seasons. Knowing more about this aspect will make windsurfing safer. There are generally two types of wind changes, one is due to air pressure and the other is due to the temperature difference. By studying the weather forecast, the windsurfer might get a certain understanding of it. In winter, a high-pressure system develops over the cold Asian continent. A large amount of cold dry air blows out from the continent and will only absorb water vapor until it reaches the ocean surface far from the land. On the east coast of Mainland China, south of 30° N, this prevalent northeast wind is called a "northeast monsoon". In winter the continental high pressure moves south. When the cold front edge arrives at the sea area near Taiwan via the East China Sea, along with it comes the northeast monsoon with substantially strong winds. During the northeast monsoon season, the north and northeast of Taiwan experience cloudy and partially rainy weather. Beginners should be aware of the northeast monsoon from October to April and the changes it brings to wind direction, wind speeds and intensity, and sea water temperature (Tsai, 2011). However, some experienced windsurfers prefer the strong waves in the northeast monsoon season as they consider them more exciting and challenging.

Windsurfing Techniques

When looking at the cross section of the wing of an aircraft, it can be seen that the wind blowing from the front of the wing will follow the shape of the wing to form two types of true wind and false wind. The wind flow passing over the wing will form a streamlined wind flow due to the influence of the wing, and the flow distance from the wind flow below the wing will increase, which will cause the flow rate to increase. That is a phenomenon that means the wind flows upward under the wing and lifts it. This kind of towing force caused by the wind is the floating force, a windsurfing sail has a similar structure to an aircraft wing, which will produce the same phenomenon. As a result, this causes the windsurfing board to move forward on the water surface (Ouadahi et al., 2016; Resende et al., 2011). If there is only the lifting force in effect, the windsurfer cannot enter the center of the wind. At this time, the floating force will only push the windsurfer and the board to the lateral direction and cannot form the driving force for advancement. In order to move forward, it is necessary to resist the force generated by the wind in the lateral direction. This depends on the dagger board and the fin on the tail of the sailboard. The resistance will offset some of the lateral pressure of the wind, but the residual lateral pressure and the forward force of the dagger board form a synergy to push the sailboard forward (Bay et al., 2018).

The angle formed between the board and the wind varies and will depend on the speed of windsurfing. The wind blows from above the sailboard will follow the wind from the mast, and the wind will flow out along the sail facing the rear (van Delden et al., 2019). The faster the wind speed is, the more the sailboard slides. When the windsurfer goes downwind, the windward angle of the sail is smaller. In this way, the wind receiving area of the sail is maintained, and the sliding force is continued. As the wind speeds increase, the speed of the sailboard accelerates. The vast majority of windsurfing enthusiasts participate in the sport to enjoy this feeling of wind speeds (Marchand et al., 2019; Sagayama et al., 2018).

Safety Issues of Windsurfing

Windsurfing often takes place at sea, so there are certain safety issues involved. Creatures in the sea may attack people, so it is best to take notice of the weather, the current, the water flow, the wind direction, the ocean tide, and other possible factors that may have an influence when participating in this sport. Before heading out to the sea, one should plan the schedule carefully, test the equipment thoroughly and avoid going out alone. Wearing a life jacket and other safety equipment in case accidents happen, and evaluating personal physical load is also important (Ouadahi et al., 2016). If, unfortunately, the equipment is broken resulting in an inability to sail back, the windsurfer has to remain calm because any wrong judgment the windsurfer makes as a result of panic may lead to serious consequences in the middle of the sea. No matter how close the windsurfer is to the shore, he or she should not risk leaving the sailboard and swimming back. If the waves and currents are too strong, the windsurfer can tear down the sail, lie on his or her abdomen on the board, and slowly slide toward the seashore. It is important not to leave the board as there is a danger if both the person and the whole equipment being carried away by the waves. The windsurfer's hands and legs might get chopped off if the angle is not right in the strong waves. If the waves are not too strong, the windsurfer should not worry or become nervous. As long as the windsurfer slowly paddles toward the shore, he or she will arrive sooner or later (Beelen et al., 2010).

Windsurfing Risks

Since windsurfing is a water sport, it is important to check the equipment before launching onto the water. In windsurfing, the intensity of the wind and the wideness of the waters are all key factors for safety. Prior to taking up windsurfing, it is important to know how to swim well, and to avoid the danger of panic when falling into the water (Menayo et al., 2016).

The Body Motion of Windsurfing

For beginners, getting to know the water and overcoming fear are key lesson (Bideau et al., 2010; Vogiatzis & De Vito, 2015). Wake sports belong to the physical endurance type of activity, which is dominated by aerobic endurance. Higher requirements are placed on physical fitness, and a good physical condition is the best guarantee of success. The main components of windsurfing preparedness should include physical fitness and endurance, speed, and explosive power, in order to maintain overall balance and stability under ever-changing sea conditions, athletes often need to use fine-tuning actions. This puts high demands on the muscle strength endurance and overall core ability of key parts of the human body and deep small muscle groups. This also shows that in strength training, explosive power and strength endurance, the static strength and endurance of small muscle groups are equally important (Figure 12.2).

Windsurfing is one of the most environmentally friendly sports. The equipment used is extremely simple, but the technology used in it is increasing. Windsurfing boards can be subdivided into various types. From the perspective of stability, the competitive board and the Fun Board are used for speed competition and skill competition. Beginners are best suited to larger boards which make it easier to master the essentials. The dagger board is fixed at the bottom of the sail board to reduce the curve in windsurfing. The foot trap is a reliable fulcrum for windsurfers. Beginners can start with a 3.0 square meter sail. Evaluating the wind speed, the windsurfer should choose a proper sail to match it, a small sail is highly recommended when the wind is strong (Mulder et al., 2012).

The buoyancy is preferably above 140 liters to prevent falling, and a sailboard with a dagger board is convenient for stability. The sail should be selected from a range of 4.5 square meters to 5.5 square meters. Sails of 5.5 square meters to 7.5

Figure 12.2 Windsurfing Training in the Country.

Source: Lucetta Tsai.

square meters or more have greater wind capacity (Naruhiro et al., 2008; Verlinden et al., 2013). Windsurfing is a sport usually carried out at sea and in the wind. In general, it requires 20 hours of practice to maneuver freely. The best condition for beginners is a seashore with a breeze, the wind blows across the coastline, and the waves are not too strong (Kulkarni et al., 2009). To sum up, it is necessary to master the essentials of windsurfing before fully participating in the sport. In recent years, windsurfing equipment has been greatly improved for the sake of windsurfers' safety and enjoyment.

Training for Windsurfing

Windsurfing is a sport that installs a sail on a surfboard and uses the wind to allow for a brisk high-speed gliding on the water. It can include various kinds of skill movements, such as vacating and big swing, which are exciting and interesting. Although there are many esoteric elements of windsurfing techniques, as long as the windsurfer masters the sliding techniques, he or she can simply enjoy the fun of the sport. The pleasure of this kind of sliding can only be truly experienced by those who have participated in this activity. However, practice makes perfect (Kojima et al., 2009).

Conclusions

Tourism development is widely considered to improve the local economy. Hritz and Ross (2010) found that sport tourism has social benefits, environmental benefits, economic benefits, and general negative impacts. Social and economic benefits were strong predictors for support for further sport tourism development revealing a strong identification with the advantages of sport tourism in a city, such as an increased cultural identity and social interaction opportunities. Chang et al. (2010) analyze marine sport tourism's impact on the development of Taiwan. According to the study, most residents recognized that marine sport tourism has a positive impact, on local leisure and recreation opportunities.

Regarding the development of marine sport tourism, windsurfing is easy to access and relatively inexpensive. It is a watersport that almost anyone can learn, is environmentally friendly and safe. Moreover, it is a social sport. The windsurfer can choose to do it either with friends or alone. To take the first step in learning how to windsurf, one should go to a spot where there is enough water with constant wind. Nevertheless, in terms of windsurfing equipment, including wetsuit, board and sail, the windsurfer can start by spending a small amount of money. Once the windsurfer has become acquainted with the necessary skills, he or she may choose to upgrade the equipment gradually. The only people who should not windsurf are children who weigh less than 30 kg and people who cannot swim (Tsai, 2011). However, a life vest could help to prevent the participant from drowning. Windsurfing is considered an extreme sport; however, if the beginner bears safety issues in mind, accidents can easily be avoided. This sport requires no fuel, makes no noise, and does no harm to the environment. Everyone is willing to share their knowledge with the beginner when someone is having trouble in windsurfing.

A current trend that has appeared lately is that young children start with basic freeride, moving on to radical freestyle. When they are no longer "as young as they used to be" they move on, but experience is just as important as physical performance. To windsurf, the participant does not rely on other people unless they are giving him or her a ride to the beach in their car. In some sports, like team sports, the player might require others to join the activity, however, it would not be necessary for windsurfing. The ancient Greek doctor Hippocrates once said, "Sunshine, air, water, and sports are the source of life and health." Outdoor activities provide human beings with the opportunity to get close to nature and experience sport tourism (Wäsche et al., 2013). Windsurfing has created a unique opportunity to meet new friends, enjoy watersports, and also to better understand the world in which we live.

References

Bay, J., Bojsen-Moller, J., & Nordsborg, N. B. (2018). Reliable and sensitive physical testing of elite trapeze sailors. *Scandinavia Journal of Medicine & Science in Sports, 28*(3), 919–927.

Beelen, M., Burke, L. M., Gibala, M. J., & van Loon, L. J. (2010). Nutritional strategies to promote postexercise recovery. *International Journal of Sport Nutrition and Exercise Metabolism, 20*, 515–532.

Bideau, B., Kulpa, R., Vignais, N., Brault, S., Multon, F., & Craig, C. (2010). Using virtual reality to analyze sports performance. *IEEE Computer Graphics and Applications, 30*(2), 14–21.

Castagna, O., Brisswalter, J., Lacour, J. R., & Vogiatzis, I. (2008). Physiological demands of different sailing techniques of the new Olympic windsurfing class. *European Journal of Applied Physiology, 104*(6), 1061–1067.

Chang, H.-M. (2009). Study of tourist cognized on costal sport tourism attractions, travel experiences, perceived values, and behavioral intension. *Journal of Leisure and Recreation Industry Management, 2*(3), 31–51.

Chang, H.-M., Lin, J.-S., Chen, C.-W., & Chan, C.-H. (2010). A study of residents recognized on costal sport tourism developing impact in Penghu. *Leisure Study, 1*(4), 19–43.

Ghiani, G., Magnani, S., Doneddu, A., Sainas, G., Pinna, V., & Caboi, M. (2018). Case study: Physical capacity and nutritional status before and after a single-handed yacht race. *International Journal of Sport Nutrition and Exercise Metabolism, 28*(5), 558–563.

Hori, N., Newton, R. U., Andrews, W. A., Kawamori, N., McGuigan, M. R., & Nosaka, K. (2008). Does performance of hang power clean differentiate performance of jumping, sprinting, and changing of direction? *Journal of Strength & Conditioning Research, 22*, 412–418.

Hritz, N., & Ross, C. (2010). The perceived impacts of sport tourism: An urban community host perceptive. *Journal of Sport Tourism, 24*, 119–138.

Jouke, C. V., Fabian, A. M., Joris, S. V., Anna, D. J., Darina, K., Zsuzsa, N., ... Paul, S. (2013). Enhancement of presence in a virtual sailing environment through localized wind simulation. *Procedia Engineering, 60*, 435–441.

Khadarooa, J., & Seetanah, B. (2008). The role of transport infrastructure in international tourism development: A gravity model approach. *Tourism Management, 29*, 831–840.

Kojima, Y., Hashimoto, Y., & Kajimoto, H. (2009). A novel wearable device to present localized sensation of wind. *Proceedings of the International conference on advances in computer entertainment technology*, New York.

Kulkarni, S., Fisher, C., Pardyjak, E., Minor, M., & Hollerbach, J. (2009). Wind display device for locomotion interface in a virtual environment. *World haptics third joint euro haptics conference and symposium on haptic interfaces for virtual environment and teleoperator systems* (pp. 184–189). Salt Lake City, UT. doi: 10.1109/WHC.2009.4810855.

Marchand, W. R., Klinger, W., Block, K., VerMerris, S., Hermann, T., Johnson, C., ... Yabko, B. (2019). Mindfulness training plus nature exposure for veterans with psychiatric and substance use disorders: A model intervention. *International Journal of Environment Research and Public Health, 16*(23), 4726.

Menayo, R., Manzanares, A., Segado, F., & Martínez, M. (2016). Relationship between amount of variability in eye motion and performance in simulated sailing. *European Journal of Human Movement, 36*, 104–115.

Mulder, F. A., Verlinden, J. C., & Dukalski, R. R. (2012). The effect of motion on presence during virtual sailing for advanced training. *Proceedings of the ISPR presence live conference*, October 24–26, Philadelphia, PA.

Naruhiro, H., Robert, U. N., Warren, A. A., Naoki, K., Michael, R. M., & Kazunori, N. (2008). Does performance of hang power clean differentiate performance of jumping, sprinting, and changing of direction? *Journal of Strength & Conditioning Research, 22*, 412–418.

Nathanson, A. (2019). Sailing injuries: A review of the literature. *Rhone Island Medical Journal, 102*(1), 23–27.

Nazim, O., Amina, A., Noureddine, A., & Mohamed, A. L. (2014). Windsurf ergometer for sail pumping analysis and mechanical power measurement. *Procedia Engineering, 72*, 249–254.

Nikolić, N., Nilson, R., Briggs, S., Ulven, A. J., Tveten, A., Forbes, S., ... Berbist, R. (2019). A medical support in offshore racing-workshop on medical support for offshore yacht races, Telemedical Advice Service (TMAS), 1–2 December 2018, London, UK. *International Maritime Health, 70*(1), 27–41.

Ouadahi, N., Chadli, S., Ababou, A., & Ababou, N. (2016). A simulator dedicated to strengthening exercises for windsurfers. *Procedia Engineering, 147*, 532–537.

Resende, N. M., de Magalhaes Neto, A. M., Bachini, F., de Castro, L. E., Bassini, A., & Cameron, L. C. (2011). Metabolic changes during a field experiment in a world-class windsurfing athlete: A trial with multivariate analyses. *OMICS: A Journal of Integrative Biology, 15*(10), 695–704.

Sagayama, H., Toguchi, M., Yasukata, J., Yonaha, K., Higaki, Y., & Tanaka, H. (2018). Total energy expenditure, physical activity level, and water turnover of collegiate dinghy sailors in a training camp. *International Journal of Sport Nutrition and Exercise Metabolism, 29*(4), 350–353.

Tsai, C. T. L. (2011). Preliminary analysis of keelboat sailing in the case of Taiwan. *Journal of Taiwan Aquatic Sport, 2*, 148–156.

Van Delden, M., Bongers, C. C. W. G., Broekens, D., Daanen, H. A. M., & Eijsvogels, T. M. H. (2019). Thermoregulatory burden of elite sailing athletes during exercise in the heat: A pilot study. *Temperature, 6*(1), 66–76.

Verlinden, J. C., Mulder, F. A., Vergeest, J. S., de Jonge, A., Krutiy, D., Nagy, Z., ... Schouten, P. (2013). Enhancement of presence in virtual sailing environment through localized wind simulation. *Procedia Engineering, 60*, 435–441.

Vogiatzis, I., & De Vito, G. (2015). Physiological assessment of Olympic windsurfers. *European Journal of Sport Science, 15*(3), 228–234.

Wäsche, H., Dickson, D., & Woll, A. (2013). Quality in regional sports tourism: A network approach to strategic quality management. *Journal of Sport Tourism, 18*(2), 81–97.

Zalaudek, I., Conforti, C., Corneli, P., Jurakic Toncic, R., di Meo, N., Antonietta Pizzichetta, M., … Curiel-Lewandrows, C. (2019). Sun-protection and sun-exposure habits among sailors: Results of the 2018 world sailing race "Barcolana" skin cancer prevention campaign. *Journal of the European Academy of Dermatology and Venereology, 34*(2), 412–418.

Part IV

Sport Tourist and Benefits Management

Photo: Miklos Banhidi

13 Managing Sport Tourists

A Case Study from Hungary

Miklos Banhidi and Tamas Laczko

Introduction

There is an increasing number of sport tourists around the world which means more and more people travel to visit sport-related events, or to practice different sport activities in specific destinations. Although Hungary is mainly focused on offering health and cultural tourism, the last few decades have seen an increase in sport-related tourism. This includes sport as a strategic sector, the increasing public participation in health-related sports, increasing domestic sports spending, and an increase in the number of internationally inspired sport events in recent years (Banhidi et al., 2014).

When people are traveling, most of them change their daily routines to focus on their travel expectations. Usually, they take more time for sleeping, eating, and relaxing. But among these routines, many active tourists seek opportunities to practice their own sports or try to participate in new challenging programs.

We think that a sport tourism destination can succeed if active sport tourists' needs and expectations are generally met. For this, providers' knowledge of policy implications is needed.

In previous decades, several national and international sport travels were evaluated by Hungarian researchers (Laczko & Banhidi, 2016). Detailed environmental analysis and evaluation of tourist feedback revealed what is needed to create an optimal sport tourism destination. During the holidays, active travelers are mainly focused on the sport activity itself, but the benefits were the main motivators, such as having fun, developing skills, and access to sports-related services, friendly sports nutrition, and sport medical help. Organizers previously and incorrectly assumed that young sport tourists do not care about the quality of services. Providing travelers with knowledgeable professionals will always be the key of the qualitative product.

From a sport science point of view, sport tourists will do anything to find facilities and services that contribute to their sport performance or joy in the activity. To find the influencing factors, a large-scale survey was administered to Hungarian adults.

Therefore, this chapter explores and identifies the traveling habits of Hungarian tourists with a particular focus on sport tourism. Specifically, the study attempts to

DOI: 10.4324/9781003476658-17

answer these research questions: (1) what the traveling motives of Hungarians are, (2) how often do they travel, (3) what kind of sport equipment do they take along, and (4) what kind of feelings do they return with after holiday.

Literature Review

Sport sociologists have repeatedly shown that sport is still stratified based on gender, class, and race, among others. Male, affluent, and college-educated are characteristics that describe the typical active sport tourist (Gibson, 1998).

In previous studies, motivation, involvement, and loyalty were examined among recreational skiers. Motivation was measured with an adjusted version of the Recreational Experience Preference (REP) scale with the three-dimensional model (attraction, centrality, and self-expression), and loyalty was measured with an intention scale. The results revealed seven dimensions: escape, social recognition, enjoying nature, excitement/risk, socialization, skill development, and achievement (Manfredo et al., 1996). A cluster analysis was used to categorize these dimensions into four segments labeled – novice, multiple-interest, naturalist, and enthusiast. Analysis of variance revealed differences among the four groups in both loyalty and the three involvement sub-scales. As a general trend, the fourth segment (enthusiast) had statistically higher scores than all other groups (Alexandris et al., 2009).

In a report on sport tourism marketing, using a qualitative research methodology, Chen (2006) utilized the Zaltman Metaphor Elicitation Technique (ZMET) because of its sophisticated imaging abilities to elicit both consumers' spoken and tacit thoughts and feelings. The findings suggest a conceptual model that combines exchange theories with essential and necessary dispositional variables to explain sport tourists' loyalty development processes. The model contains four antecedents of loyalty with 'exchange relationships' as the moderator and 'trust' as a precursor. The concept of social and resource exchanges reflecting the long-term nature of consumer consumption processes enables the model to capture the complex and dynamic relationships in the loyalty development processes. It also rises to the marketing challenge of building long-term consumer relationships. Sport tourists' loyalty, therefore, might be improved by maximizing trust and several or all these antecedents through customer relationship building using social and resource exchanges (Chen, 2006).

The economic perspective of sport consumers' behavior became an important topic among researchers. The focus was to develop a deeper understanding of tourists' perceptions and experiences of tourism destinations as one method to ensure the location's long-term sustainability (Papadimitriou & Gibson, 2008). Five benefits that were discovered include socializing, sport experience, excitement, enrichment, relaxation, along with three image dimensions attractions, environment, and outdoor activity. The pre- and post-trip analysis largely revealed congruity between pre-trip expectations and post-trip satisfactions.

A conceptual framework drawing upon role theory, life span, and family life cycle models, motivation theories, and the effects of social structure (gender, race, and social class) can be used to further understand sport tourism behavior (Gibson, 2006).

Consequently, sport tourism can provide serious leisure (Stebbins, 1992) participants with: (1) a way to construct and/or confirm one's leisure identity, (2) a time and place to interact with others sharing the ethos of the activity, (3) a time and place to parade and celebrate a valued social identity, (4) a way to further one's leisure 'career', and (5) a way to signal one's career stage. By understanding the nexus of serious leisure, social identity, and subculture, they were better able to describe and explain participation in what we term 'serious sport tourism' (Green & Jones, 2005).

Some researchers have been focused on how to attract more active tourists. Results of a research project suggest active sport tourists' event image perceptions are related to the themes with emotional aspects, being that they were more dominant during the post-event recall (Kaplanidou, 2010).

Sport tourism is definitely a popular tourism product (Gibson, 1998). The reason is because sport can be a good tool to have fun, and to fulfil many tourist expectations. Sport can be a good tool to have active rest, relaxation, to enjoy spending time with family, and friends (McCarthy et al., 2008), or experience challenging activities, develop special skills, or discover themselves (Holt, 2016). In an Asian study, travel motivations were measured among Chinese tourists, who mainly travel to destinations where they can relax (25%), have a good time with friends or family (22%), and seek a better understanding of foreign cultures (17%) (Toh, 2017).

Moreover, sport tourists are different from each other based on their motives. In a review of the literature, Smith (1986) and Gibson (2003) classify sport tourism into the following categories:

- Professional sport tourists (athletes)
- Commercial sport consumers (ad hoc)
- Mass sport tourists (spontaneous)
- Health-focused (conscious lifestyle)
- Extreme (explorers – seeking adventure)
- Off-beat, unusual (doing something unique)

Primarily, professional sport tourists are athletes who travel to competitions, representing their country or team. They are focused on results, such as how to win or reach greater results, requiring them to maintain a strict program and less spare time (Gilbert, 2014).

Commercial sports tourists are sports consumers who buy from the tourism provider's choices. Their behavior is influenced by fashion and the results of top-level athletes. They like wearing fashionable clothes and buying souvenirs (Green & Jones, 2005).

Mass sport tourists refer to travelers who don't plan to do sport activities. They do it only ad hoc, mostly by following others. Their main motives are connected to entertainment with bigger groups of people.

Near tourism destinations, there are more and more travelers who design their daily programs following the healthy lifestyle instructions (Kwak, 2011). They plan regular physical activities and eat to make up for the lost energy (Table 13.1).

Table 13.1 Changes of Travel Motivation Level among Hungarians (1,000 trips)

Purpose of traveling	2013	2014	2015	2016	2017
Leisure tourism	311,099	371,583	409,614	464,792	516,937
Visiting relatives	80,536 (26%)	93,429 (25.1%)	103,809 (25.3%)	106,626 (25.3%)	97,915 (18.9%)
Sport, health, and wellness	791 (0.25%)	1,064 (0.29%)	1,482 (1.4%)	1,066 (0.34%)	3,037 (0.59%)
Business tourism	62,066	58,187	75,385	80,602	74,617
Conference	5,691 (9.2%)	7,473 (12.8%)	8,043 (7.6%)	6,142 (7.6%)	5,567 (7.4%)
Shopping	2,885 (4.6%)	5,385 (9.3%)	5,394 (7.2%)	5,905 (7.3%)	6,245 (8.4%)
Work	38,700 (62.4%)	32,500 (55.9%)	35,007 (46.4%)	43,218 (53.6%)	54,511 (73.1%)
Study and others	7,633 (12.3%)	11,152 (19.2%)	4,672 (6.2%)	7,349 (9.1%)	5,175 (6.9%)
SUM	422,382	478,807	530,072	601,865	657,485

Source: KSH (2017).

Extreme sport tourists are usually dependent on the natural environment, which challenges the human capacity. Places with extreme climate, high altitude, or deep water remain target destinations for these extreme activities.

Off-beat sport tourists are travelers who want to learn new activities, usually for the first time. They usually need help from tourism and sport providers. Off-beat sport tourists do physical activities just for the sake of doing something unusual. They don't care much about the activity itself. For them, the priority is to have fun, become acquainted and work with others.

However, Hungary is not fortunate enough to be endowed with spectacular features, either natural or man-made. This shortcoming is compensated for by its geographical situation and the sheer number of its modest attractions (HNTO, 2018). In addition, Hungary is situated in the center of Europe, which has the largest amount of tourist traffic in the world as well as a strong tourism reputation of hosting and traveling.

In Hungary, the National Statistical Office does not publish a category that clearly identifies sport tourism. There are several categories that are distinct, partly involving active or passive sport tourism, but it is difficult to draw adequate conclusions from such information (KSH, 2017). Previous surveys carried out among the Hungarian adult population reveal that the most popular holiday activity is hiking (63%), with high numbers of people attracted to cycling (9%), and young people showing a special interest in water sports (14%) (HNTO, 2018).

Fortunately, the unfavorable picture of Hungarian tourism is positively influenced by the increase in travel expenses and the strong turnover of commercial accommodation in recent years. In 2016, spending on multi-day domestic travel was only 6.2% higher than the total spending in 2008, but more than 30% of the 2009 travel expenditures (KSH, 2017). Considering the general inflation rate over this time period, expenditure from 2009 to 2016 exceeded the rate of inflation, which cannot be said for the 2008–2016 period. As a result of the decreasing number of trips and increasing expenses, the average total daily expenditure per capita increased by 55.9% over the period under review. This data reveals that the structure of domestic tourism is constantly changing. Reduced trip motivators such as relatives and acquaintances or hobby-type work have lost their previous role, while the rates of travel with higher average spending (such as holidays, visits to cultural and sporting events, or health tourism trips) have increased over the last nine years. In addition, although a significant proportion (50%) of the population are excluded from multi-day domestic tourism year after year, those who visit domestic sites yearly spend more and more money on these types of trips.

Goals and Methods

For this chapter, both quantitative and qualitative methodology was adopted (mixed methods). For the analysis, national statistical datasets were included, motivation scales involving traveler loyalty were examined to develop a typology of user segments. An extensive review of the literature was also performed with a focus on travel behavior analysis. We also performed a short comparative analysis on what travelers expect during holidays and how they spend their time at the destinations.

The main goal of the study was to get more information on travelers who focus on sports (hiking, cycling, and water sports) during their holiday. The results can deliver important information for tourism service providers and disclose how they can offer better service for sport tourists with different needs.

Also, a post-trip design was used to explore the desired benefits, satisfaction, and the traveling habits of Hungarian sport tourists visiting several tourism destinations. The trials were Hungarian adults (N = 3,961). A self-administered questionnaire was completed by 43% male and 57% females living in North-Western Hungary at the time of the study. After recording the data, we split the trials/research subjects into two groups: (1) those who were practicing different sports – hiking, biking, and water sports – for at least an hour per day during their holidays; and (2) those who were not engaged in sport activities. For the evaluation phase, descriptive and comparative analyses were performed using Statistica 11 software.

Results

Travel Motivations and Expenditure of Hungarians

In Hungary, every age group partakes in recreational holidays. In 2017, more than 500 million domestic leisure-related travels were registered, which is 51 trips per citizen. Tourism offers rich experiences and has long-standing traditions in Hungary (see Figure 13.1). The country has more than 21,000 km of marked hiking trails, coordinated by 322 associations. One of the most popular sites is the oldest

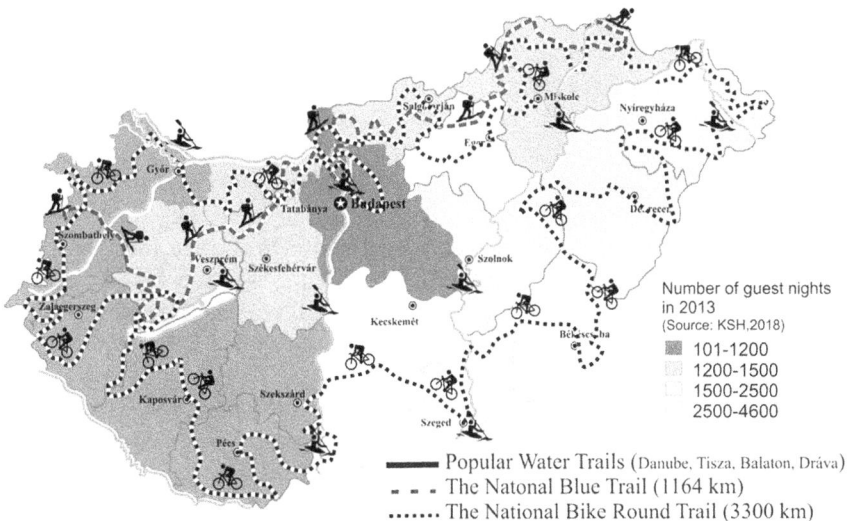

Figure 13.1 National Sport Touristic Trails and Regional Distribution of Guest Nights in Hungary.

long-distance path in Europe, the National Blue Trail, which was established almost 80 years ago in 1938. In this small country, wild camping is allowed only with a few limited conditions. Also, cycling belongs to the most popular tourist sport activities for Hungarians. To support the growing market, a long-distance bike trail was established across the country in 1977, which has reached 3,300 km. The tourism of Lake Balaton had a great impact by establishing the 206 km cycling route since 1993. It runs through 138 settlements, offering 456 tourism attractions (KMSZ, 2008).

Based on the national statistical data from previous years (2013–2017), most of the leisure trips were motivated by visiting relatives and just few of them were health- and sport-related travels (see Table 13.1). Among the travelers, young people prefer active holidays (e.g., hiking, cycling, and water sports), families opt for destinations where there are facilities for children (e.g., child-friendly hotels and swimming pools with special attractions), and members of the third generation look mainly for relaxation in their recreational holidays (KSH, 2017).

Although the number of sports travels are not increasing very quickly, the economic potential is becoming dominant. The Hungarian Central Statistical Office published in the annual report that in 2016, Hungarians spent 1.18 billion Hungarian Forint (HUF) on sport and wellness services and devices during their domestic holidays, accounting for 0.38% of total travel expenses (KSH, 2017). This amount is 32% higher than in 2009, but by examining the data for that period, it can be said that sport spending has been fairly hectic in recent years. Despite the stagnation of daily average sport spending per capita in the period 2013–2016, overall, it increased by 62%, rising significantly over the period 2008–2016 (in 2016 alone = 20.16 HUF/day/person) (Table 13.2).

Looking at the geographic distribution of sports expenditure related to domestic tourism, the greatest amount of sports spending was at Lake Balaton (38%) and in the Budapest-Central Danube Region tourism districts (14.7%). Together, these expenditures accounted for more than half of the national total spending.

Some of the effects of demographic, social, and economic factors on domestic sport tourism expenses are explained by the examination of general sporting habits associated with journeys during the 2009–2016 period. These trends include men, younger generations, or those who work locally, spending more on their travels than women, older people, or inactive groups. On average, men spent 80% on sports, just like the women. Overall, people ages 25–44 and 44–65 pay most for the use of sport services during their domestic travel. The average daily cost of the 25–44 age group is 15% higher than ages 15–24 and 54% for the 44–65 age group.

Like the influence factors so far, the impact of school education has long been known for its effect on general sporting habits and sport consumption. This impact is also evident in the amount of sport spending associated with travel. The only apparent contradiction is that most of the sport spending associated with domestic travel (40%) comes from secondary school travelers since they form the far-flanked group of people on a national multiple day trip. However, in the daily average amount spent on sports, the difference in the eight-year period is that the

Table 13.2 Changes of Sports-Related Spending during Multiple Days Domestic Trips for 2009–2016

Year	Spending for sport (1,000 HUF)			Daily expenditure on sport (1,000 HUF)		
	All multiple-days domestic travels	Rest, entertainment	Cultural and sport events	All multiple-day domestic travels	Rest, entertainment	Visiting cultural and sport events
2009	894,943	644,129	19,223	12.4	18.8	34.5
2010	1,187,144	924,607	14,968	16.4	27.4	31.9
2011	838,389	685,660	21,162	11.2	20.4	34.1
2012	957,489	773,696	33,909	13.5	24.8	51.1
2013	1,224,391	897,105	154,385	20.1	32.5	262.5
2014	1,241,097	905,997	123,087	20.2	33	192.3
2015	1,267,048	742,431	224,840	20.4	26.1	252.5
2016	1,188,660	807,169	215,386	20.2	27.9	229.2

Source: KSH (2017).

average expenditure of people with a post-secondary education is the highest, and the lowest for primary and non-graduates.

As for the size of the place of residence for domestic travelers and the organized trips, we see slightly different and less stable tendencies in travel related to sport spending as we assumed in the average sporting habits. Based on the categorical examination of the domestic travelers' place of residence in sport, it can be said that the phenomenon of "settlement slope," known and verified in the Hungarian social sciences, does not exist in this area. In 2016, most of the money spent on domestic sport travels produce more than one-third (37.5%) of total spending. It should be noted that the lowest average daily sport spending was typically found in middle-sized cities and Budapest, while for the other small settlements category, these amounts exceeded the national average. The sport expenditures of the capital residents have declined significantly in recent years (by almost a quarter of 2016). In the earlier periods, Budapest was the group with the highest spending. Cities with more than 100,000 inhabitants tended to have the lowest total and average spending on sport during their 2009–2016 domestic voyages.

In Hungary, only 1.4% of all holiday trips are organized by travel agencies or other travel agents. As a result, the amount spent on sport during travel is primarily related to private travel (94.2% of the total expenditure), and it is important to point out that the average daily spending in sports is far behind the expense of trips managed by travel companies. On average, expenditure for sports spending on domestic travels that are organized by travel agencies in 2010 was 4.4 times higher than the average daily expenses of travelers planning their own trips. Despite this fact, in 2009–2010, the average sports spending on travel agents' travels were even lower than for individually organized trips. This trend turned in 2011, and in the following period, the difference between the participants in organized trips has become increasingly significant year after year.

Behavior of Sport Tourists

In our case study, we analyzed Hungarian adult's traveling habits with a focus on how they spend their holidays, for people who practice everyday sports and those who do not. We also explore their motives for traveling, how often they travel, what kind of sport equipment they take, and what feelings they experience after returning from holiday. We also made a comparative analysis that focuses on the difference between all travelers, and people who practice sports for more than an hour during their holiday. At the beginning, we thought sport tourists were only or mostly interested in sport activities and focused less on other issues. The results revealed a different picture (see Figure 13.2). The results regarding the priorities of active tourists are much different than that of passive tourists. Adults who spent at least one hour doing sport were also paying attention to other issues, meaning they spend their holiday more consciously. On the exterior, they appear to take holidays for the company of others, but many of them concentrate on mental and physical rest and seek opportunities to visit sights and try adventure activities.

The average values of all passive tourists were lower than active tourists. This means that, generally, adults don't care too much about details during their holiday.

Priorities During Holidays (%)
N = 1,555

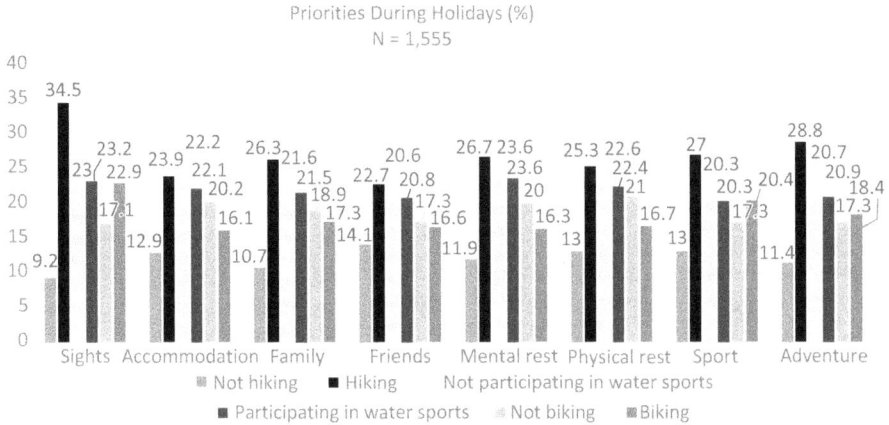

Figure 13.2 Priorities of Interviewed Active and Passive Travelers.

Note: The bars in this graph match the keys when read from left to right in order.

Traveling Frequency of Active and Passive Travelers (%)
N=1555

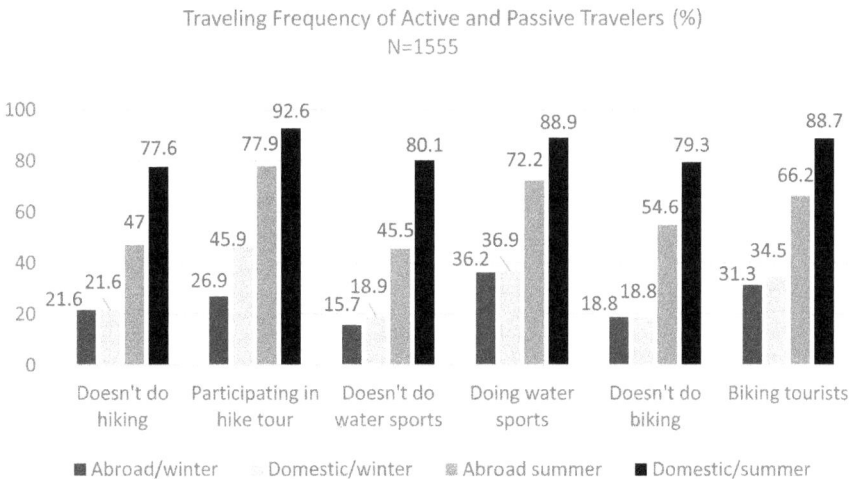

Figure 13.3 Traveling Frequency of Adults Who Engage in Sports at Least 1 Hour during Holiday.

Note: The bars in this graph match the keys when read from left to right in order.

For few of them, important factors included accommodation or spending time with family or friends. More interestingly, every third interviewee had interest to participating in sports.

We were also curious to discover if active sport tourists traveled more than other adults. The results show that they prefer traveling every year, mainly to domestic destinations and especially during summertime (57.7%) (see Figure 13.3). Few of them chose winter vacations at home or traveling abroad, proving that seasonality plays a big role in the frequency of travel. In addition, our survey reveals a significant difference between the active and non-active tourists. A higher percentage of active sport tourists travel in the summer and winter time, taking preference to domestic destinations.

In the questionnaire, we asked participants what kind of sport equipment they carry for their multi-day holiday activities to discover whether active travelers have them more than other adults. The results show significant differences between active and passive tourists (see Figure 13.4). Many of the respondents (85.3%) stated that they like taking a bike while traveling with the hope to have an opportunity to use it. In an earlier study, we found that 98% of Hungarian families have at least one bike and use it at least twice per week (Banhidi, 2007). The high percentages are understandable, as bike tourism in Hungary has become one of the most popular sport tourism activities. As a result, travelers take their bikes with them when they travel because there are still not enough bike rental opportunities in tourist destinations. Other favorite activities for Hungarian travelers are ball and racquet games along with beach sports in the plentiful beach areas. The survey confirmed its popularity; 33.3–70.8% of the passive and 70.9–91.8% of the active travelers (especially water sport tourists and bikers) reported that they always had a ball with them during a holiday trip. We can see that among tourists who practice water sport activities, 91.8% of them carry water sport equipment, and just 4.5% of the passive tourists have water sport equipment with them.

These days, waterfront areas in Hungary offer many water sport rental services, which give everyone opportunities to engage in these sporting activities. Obstacles to use the equipment are the lack of knowledge, guidance, and higher costs compared to other equipment. It seems to be that water sport tourists like to carry their own equipment, even if its transportation is not easy. The general feedback

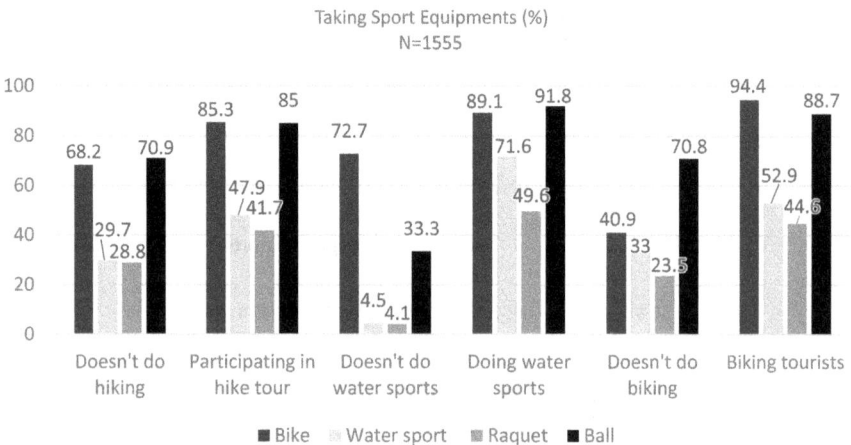

Figure 13.4 Percentage of Adults Who Took Sport Equipment with Them during a Holiday Trip.

Note: The bars in this graph match the keys when read from left to right in order.

of Hungarians is that tourism services don't always offer equipment that meet the tourist's expectation (Hennessey et al., 2002). This then explains why most of them feel the need to bring with them their own equipment.

Conclusion

The overarching conclusion is that the growing sport tourism market should be analyzed more from a global perspective. The main questions related to our topic are not easy to answer. We know that different types of sport tourists exist because they come from different backgrounds, different cultures, lifestyles, and they have different skills and experiences. Even if they practice the same sport, their expectations and attitudes are not always similar. Most tourists travel because they like to relax, rest, or spend time with family and friends.

The results from our questionnaires can be summarized as follows:

- There are significant differences between tourists who engaged in sport activities and those who did not.
- Tourists who practice sports, whom we refer to as active sport tourists, spent a significant amount of time doing something active, often using sport services.
- Sport tourists from our survey are more motivated to spend time with their families and friends and to participate in adventurous activities.

When we speak on sport tourists, we suggest that physical activity can help them reach a higher quality of time at destinations. Tourism destinations should get prepared to host sport tourists. In sum, tourism destinations, including service providers, need to be prepared to meet the needs of this growing trend of sport tourists. Understanding their needs and expectations can help tourism service provider tailor their services to meet such expectations. This then can help boost the experience of sport tourists, which in turn could be beneficial to the tourism industry.

References

Alexandris, K., Kouthouris, C., Funk, D., & Giovani, C. (2009). Segmenting winter sport tourists by motivation: The case of recreational skiers. *Journal of Hospitality Marketing & Management, 18*(5), 480–499.

Banhidi, M. (2007). Bike, Tourism, Science. [Kerékpár, turizmus, sporttudomány]. In E. Kiss (Ed.), *Pedagógián innen és túl: Zsolnai József 70. Születésnapjára* (pp. 675–683). p. 941 Pápa; Pécs: Pannon University BTK; University of Pécs.

Banhidi, M., Simonek, J., & Dobay, B. (2014). Prefered sport tourism destinations of Slovak and Hungarian tourists. *Sport Science, 7*(2), 73–76.

Chen, P. (2006). Sport tourists' loyalty: A conceptual model. *Journal of Sport Tourism, 11*(3–4), 201–237.

Gibson, H. (1998). Active sport tourism: Who participates? *Leisure Studies, 17*(2), 155–170.

Gibson, H. (2006). Cultures, commerce, media, politics. Towards an understanding of 'why sport tourists do what they do'. *Sport in Society, 8*(2), 198–217.

Gibson, H. J. (2003). Sport tourism: An introduction to the special issue. *Journal of Sport Management, 17*(3), 205–213.

Gilbert, G. (2014). *Sporting championship. Tour de France.* Weigl Publishing.

Green, B. C., & Jones, I. (2005). Serious leisure, social identity and sport tourism. *Sport in Society, 8*(2), 162–181.

Hennessey, T., Morgan, S. J., Elliot J. P., Offner, P. J., & Ferrari, J. D. (2002). Helmet availability at skiing and snowboarding rental shops: A survey of Colorado ski resort rental practices. *American Journal Preventive Medicine, 22*(2), 110–112.

Holt, N. (2016). *Positive youth development through sport.* Routledge.

HNTO (Hungary National Tourist Office). (2018). *Tourism in Hungary 1990–2002.* Hungarian National Tourism Office. Retrieved May 4, 2018 from https://visit-hungary.com/download.php?docID=2018

Kaplanidou, K. (2010). Active sport tourists: Sport event image considerations. *Tourism Analysis, 15*(3), 381–386.

KMSZ (Hungarian Cycling Federation). (2008). *A Balaton régió kerékpáros-turisztikai stratégiája és a bringakörút projekt előkészítő dokumentuma [The bike tourism strategy of Lake Balaton region].* Balatoni Integrációs Kht. Retrieved April 15, 2018 from http://www.terport.hu/webfm_send/693 15/4/2018

KSH (Hungarian Central Statistical Office). (2017). *A külföldről hazatérő magyarok utazásai.* Hungarian Central Statistical Office. Retrieved May 4, 2018, from http://statinfo.ksh.hu/Statinfo/haViewer.jsp

Kwak, D. H., Kim Y. K., & Hirt, E. R. (2011). Exploring the role of emotions on sport consumers' behavioral and cognitive responses to marketing stimuli. *European Sport Management Quarterly, 11*(3), 225–250.

Laczko, T., & Banhidi M. (2016). *Sport- és egészségturizmus alapjai [Basics of sport and health tourism].* University of Pecs Publisher.

Manfredo, M., Driver, B., & Tarrant, M. (1996). Measuring leisure motivation: A meta-analysis of the recreation experience preference scales. *Journal of Leisure Research, 28,* 188–213.

Papadimitriou, D., & Gibson, H. (2008). Benefits sought and realised by active mountain sport tourists in Epirus, Greece: Pre- and post-trip analysis. *Journal of Sport & Tourism, 13*(1), 37–60.

Smith, R. V. (1986). The potential of tourism as a factor in Ohio's economic future. In J. Frazier, B. Epstein, & J. Langowski (Eds.), *Papers and proceedings of applied geography conferences,* (Vol. 9, pp. 143–153). State University of New York, Binghamton.

Stebbins, R. (1992). *Amateurs, professionals and serious leisure.* McGill-Queens University.

Toh, V. (2017). *Sabre unveils evolving behaviours and motivations of today's Chinese travellers.* Retrieved October 9, 2017, from https://www.sabre.com/locations/apac/releases/sabre-unveils-evolving-behaviours-and-motivations-of-todays-chinese-travellers/

14 Human Benefits

Socio-Cultural, Mental, Physical, and Learning Outcomes

Bijen Filiz and Giyasettin Demirhan

Introduction

Tourism is one of the most important industries that directly and indirectly impact people's physical, social, and psychological needs. One of the main reasons countries are closely interested in tourism is its ability to contribute to the population's welfare. In this context, focusing on sustainable tourism activities as alternative products that can be developed locally is important to develop the region's welfare. In doing so, it is necessary to mobilize appropriate alternative tourism types that will provide cultural and social interaction. Sports tourism is one of the most suitable for these criteria.

Ross (2001, p. 3) defined sports tourism as travel and experiences to do or watch sporting activities. Thus, sports tourism includes traveling, watching, active participation, and gaining experience. Traveling for sports, playing, or watching sports in places far from one's origin, allows people to acquire physical, mental, and social benefits, learn new things, interact with new cultures, and gain new experiences. In fact, being open to innovations, curious to learn, passionate for adrenaline, desiring to escape routine environments, and the desire and courage to test physical strength, encourages people to participate in the sports environment in new and different cultural textures. Sport tourists who participate in the sports environment are satisfied with their physical strength, accumulate memories, interact with people and cultures, return to normal life with new things, use part of their gains in their vital skills, and happily continue their lives by renewing themselves. In this context, social, physical, mental, and learning outputs of sports tourism seem to have many positive effects on the individuals.

The Effects of Sports Tourism on People

Sports tourism has a social qualification in terms of the values it has. Social effects mean that tourism has directly or indirectly effected existing social values, personal behaviors, family relationships, the concept of security, ethical rules, artisan activities, traditional demonstrations, and social structure (Brunt & Courtney, 1999). Therefore, sports tourism can lead to changes in the socio-cultural structure that embraces people's value systems, traditional lifestyles, family relationships, and personal behaviors (Ratz, 1997).

DOI: 10.4324/9781003476658-18

Tourism is a dynamic process in terms of creating an interaction between tourists and local people (Fagence, 2003, p. 63). Therefore, tourism can be seen as a means of social change (Rızaoğlu, 2004, p. 19). Tourists may affect people with whom they communicate with their lifestyles and behaviors. Every year, thousands of people make a great deal of important changes with tourism in family structure, beliefs and values, cultural, political, and social structure (Akman, 2007, p. 134). If tourists and local people have similar characteristics in terms of economic, educational, and cultural levels, social change occurs less. On the other hand, if there are great differences in the cultures and economies of the countries, social change is more likely (Demircioğlu, 1997, p. 137).

Socio-Cultural Effects

Researchers have tried to classify the social effects of tourism. Pearce (1989) identified six basic categories in the classification of social and cultural effects: effects on population structure, occupational changes with tourism, changes in the values of society, tourism's effects on the traditional way of life, the effects of population density, and the benefits of tourists. Pizam and Milman (1986, p. 29), with the same idea, classified socio-cultural effects of tourism as demographic (population size, age), occupational (changes in the occupations), cultural (changes in tradition, worldview, and spoken language), social values (morality and others), effects of population density (infrastructure), and effects on the environment (pollution).

While classifications were made in this way, Travis (1984, p. 22) created a model that shows effects of a touristic destination in terms of socio-cultural advantages and disadvantages. According to this model, tourism can lead to socio-cultural beneficial or objectionable consequences. For example, cultural development and change, social change, positive image of the local community with tourism, improvement of health conditions with new investments, development of social facilities, education, and communication between people, were considered as positive changes. Negative changes were corruption or disappearance of local culture, changes in rules and laws, occurrence of commercial relations between local people and tourists, and change of traditional values (Akman, 2007).

As a result of social communication in tourism, changes in social values, quality of life of society, value systems, work sharing, family relations, attitudes, behaviors, and sense of entertainment can be observed in long term (Pizam & Milman, 1986, p. 13). On one hand, this change is caused by tourist culture and authentic memories. On the other hand, the fact that the local community is dependent on tourism, and the whole or a substantial part of their income is earned with tourism, also cause these changes (Sharpley, 1994, p. 58). Some of these changes (e.g., increase of income, education, employment opportunities, and developments in infrastructure and services) are perceived as positive (Lankford & Howard,1994), and others (e.g., changes in social and family values, emergence of new powerful economic groups, adaptation of customs according to tourists' needs, security risks, environmental damage, social conflicts, and traffic problems) are perceived as negative (Ap & Crompton, 1993, p. 48).

In sports tourism, one of the important products of social interaction and change, tourists from developed countries are responsible for carrying the values of their own countries. Countries that host sports tourism organizations put their facilities, staff, accommodations, businesses, and shopping centers into service for tourists. This enables a variety of social and cultural changes to take place by performing a certain amount of cultural exchange between tourists and local people. However, these social and cultural changes take place both positively and negatively (Bektaş, 2010, p. 20).

Fredline (2005, p. 268) defined positive social effects of sports tourism as pride, self-realization, entertainment opportunities, community or family cohesion, inter-cultural interaction and nationalist emotions resulting from team competition, reduction of psychological prosperity due to perceived loss of control over the local environment, and bullying. Higham (1999) defined negative social effects as crowding, infrastructure congestion, exclusion of local people due to costs, de-terioration of local lifestyle, displacement of local people, bad behaviors of sup-porters, and suppression of human rights. Studies have shown that hosting major sports events can increase the local population's sense of belonging to the commu-nity (Delamere, 2001), because organizations increase awareness of the region and culture at national and international levels and also increase the quality of life of residents (Deccio & Baloglu, 2002; Goeldner & Long, 1987). For example, more than 1.5 million people celebrated the Champs-Elysees victory of the French Team in the 1998 World Cup in Paris (Choi, 1999). This sense of community can be en-hanced by organizing cultural activities and entertainment with various organiza-tions (Pugh & Wood, 2004).

However, various terrorist attacks and terrorist threats over the world have in-creased public concern about risk perception for sports events (Hall et al., 2008; Taylor & Toohey, 2007). For example, a football match in Turkey played on Sep-tember 17, 1967, between Kayserispor and Sivasspor was a fatal event caused by violent football spectators in which 43 fans died and hundreds of people were in-jured. Another example, the Heysel Disaster, was an event in which 39 fans died because of a wall that collapsed after Liverpool fans attacked the Italians before the final match of the European Championship Cup, played in Brussels on May 29, 1985. These unfortunate events defamed sports organizations and will always be concrete events to underline the social adverse effects of sports tourism. In this context, the security issue plays an important role in the planning of large-scale sports events.

Another issue is that local people are faced with such troubles as traf-fic disruption and infrastructure construction in the formation of sports or-ganizations (Standeven & de Knop, 1999). Ritchie et al. (2009) noted that local people have generally developed supportive behaviors toward hosting sports events but are also concerned about traffic congestion, parking prob-lems, and adverse economic effects (e.g., potential increases in living costs). Event managers should provide systematic and practical solutions to reduce the negative perceptions of hosting large-scale sports organizations among residents (Kim, 2012).

Physical Effects

Gibson (1998, p. 3) defined sports tourism as "Leisure time-based travel that people leave their current place to temporarily participate in physical activities, to watch physical activities, or to visit attractions associated with physical activities". In this context, there are many types of sports participation for tourists. A sport is an activity associated with joy and fun. It also ensures that physical strength and performance are well demonstrated. For example, people who participate in sports organizations as spectators can have fun or face dramatic emotions. Those participating in active sports tourism activities can exhibit their sportive performances. Physical fitness, fun, and rules of the sports are unique characteristics of sports tourism (Argan, 2004). Nogowa et al. (1996) revealed that tourists interested in active sports concentrate on a single sports activity and perform these activities for fun, healthy living, and sporting affection.

Those who are in the category of active sports tourists are divided into two groups as elite or recreational. Those interested in recreational sports apply different sports branches to overcome stress, a classical disease of intense work environments in today's world. People in this group generally act collectively and also participate in social and cultural activities (Gürsoy, 2016).

In terms of active sports tourism, there is a wide range of activities organized in natural and artificial environments such as mountain, air, water, nature, cycling, golf, winter, cave, and orienteering tourism. These events are called hard and soft sports tourism (Garda, 2014). For example, river rafting, scuba diving, and mountain biking are considered hard activities; activities such as camping, day-tripping, animal observation, horse riding, sea-kayaking, and water-skiing are soft activities (Beedie & Hudson, 2003, p. 632). Some of these activities require individuals to be physically prepared and trained in advance.

The physical effort required by the fascinating and fragile natural environment is an incentive for sustainable tourism. In order to struggle through the tough physical conditions, mental and emotional skills, as well as physical ability must be developed. To successfully avoid risk increases the confidence of the person as well as his/her personal skill. Supplying a natural environment where people can push their physical limits will increase the attractiveness to physically active tourist activities. On the other hand, the development of physical and technical skills will ensure risks are reduced. Equalization of the limits of danger and skill creates endless opportunities in terms of tourism activities (Garda & Temizel, 2016).

The activities that require challenging physical exertion have a more difficult and faster structure. Participants prefer activities within natural environment to broaden performance and achievement limits. As a result, innovations in sporting events such as skiing, mountaineering, climbing, and rafting are being tested by professionals with superior skills who want to improve their techniques. On the other hand, amateurs want to implement these innovations and to achieve similar standards achieved by professionals (Garda & Temizel, 2016). Thus, people experience a sense of accomplishment and self-renewing from sports that require compulsive, strenuous, and physical power in different environments.

Mental Effects

People, apart from doing physical activities, participate in sports tourism activities in order to take a rest, to see a new place, or to gain new experience. This participation leads to significant results and changes on human life and mental status. Motivation to participate in sports tourism activities and decision-making are important mental processes. In addition to changes in mentality and attitudes during activities, the effects of these changes on behaviors and the reactions that people develop to adapt to activity challenges are largely linked to psychological and mental processes (Ministry of National Education [MNE], 2014).

Self-realization is at the highest level of human needs that directs human behavior. In other words, the better a human's life is, the better their development, and the better they can heal. When a person is a tourist, they expect to know and understand themself better, with a belief to explore the depths of the soul and to taste the joy of existence. A person who has reached the consciousness of realization can look at the tourism event as a healthy, conscious, and real need. They participate in tourism and exhibit the necessary characteristics according to the quality guest profile (MNE, 2014).

Participating in tourism activities is an effort to increase people's mental and physical quality of life (Griffin & Stacey, 2011). Participation in these activities seems to have indirect and direct positive effects, such as providing more happiness, healthier lives, longer lives, higher self-confidence, and life satisfaction (Sarı & Özdemir, 2014). In addition, sports made for entertainment purposes are mentally relaxing. Sport is an activity that gives people the chance to discharge themselves while offering an opportunity to alleviate their problems. Kasap and Faiz (2008, p. 24) stated that this situation contributes to positive thinking for the participants. It was noted that people participating in tourism activities experienced mental and physical healing with relief from tension and depression, purification from pressure of achievement, and less monotony in everyday life (Schmidhauser, 1989).

MacCannell (1977) noted that tourists are motivated to get rid of routine and seek authentic experiences. On the other hand, Iso-Ahola (1982) pointed out, in socio-psychological terms, that sports and tourism behavior is motivated by two forces: seek and escape. The impulse of escape is dominant in tourism behavior in terms of moving away from one's daily environment. On the other hand, tourists can choose activities that involve a sports event that meets the need for optimal arousal. Hartmann (1982) determined the impulses of tourists' "escape" as the desire to get rid of loneliness, exhaustion, and distress in everyday life, and the desire to be involved in an emotionally rewarding community. Hartmann determined the impulses of "seek" as traveling for a specific purpose, such as to take a social position, to gain strength, and to be respected. Therefore, it is seen that sports tourists are motivated by various disciplines.

Crompton (1979) stated that people may be encouraged by pushing factors such as rest, prestige, health, fitness, adventure, enjoyment, social interaction, family cohesion, and excitement and by pulling factors such as beaches, cultural events, natural scenery, and shopping. These pushing and pulling factors can work together

(McGee et al., 1996). Tourists can participate in tourism activities to escape every-day routine life and to find authentic experiences.

During sports tourism activities people may be affected by stimuli that are not accustomed. By combining one's past experiences, the tourist constantly controls the field of activity, past experiences, memories of symbols, and thoughts before the activity. In the mental process, this may cause the thoughts of tourists to change. If there are big deviations between the expectations and the activity itself, tourists may start to feel dissatisfied with the experiences (MNE, 2014). In order to avoid such disappointments, people need to be prepared realistically in sports tourism.

Learning Effects

Tourism, which is generally an activity related to taking a rest, sports, and culture, is also an important opportunity in terms of self-education and learning about the differences between communities/cultures. Crompton (1979) divided tourists' motivations for recreational holidays into two categories: socio-psychological motivations and cultural motivations. Cultural motivations include innovation and education. The desire to recognize cultural differences and to learn new things are driving people to travel to innovations and to gain new experiences. In addition, people meet an important part of their socio-psychological needs by communicating with their environment in the holiday environment.

Tourists want to visit a complex product to meet many needs at the same time. For this reason, tourism products are seen as experiential product (Goldsmith & Tsiotsou, 2012, p. 208). The qualities of socially constructed tourist experiences include a range of qualities, such as symbolism (meaning, feelings, and emotion), socialization (joining with local people and participation), seeing new places, learning new things, and making memories (Batat & Frochot, 2014). For example, watching a major sports event in a touristy city can leave a person with unforgettable memories if their team wins. In addition, sports tourism increases the intellectual enjoyment resulting from learning new things and the relationships arising from meeting with people with different cultures and characteristics (MNE, 2014).

Giving tourists the feeling of "being there" instead of "looking at" (Markwell, 2004) and the opportunity to be a "co-creator" of the experience enriches their experience. In this context, activities should be evaluated in cooperation with tourists and tourists should be encouraged to give feedback and to share their ideas to enrich the activities.

Sports tourists often make an effort to learn by trying new sports in recreational environments. For example, in Alaçatı-Turkey, sports tourists get training in windsurfing, kite surfing, and dinghy. In Antalya, guides give training to tourists before going down to the river for rafting. Tourists must be trained for a certain amount of time in order to use professional canoes. Training is given in advance about the material used for rock climbing techniques and how to climb. Tourists develop positive attitudes about behaviors such as enjoyment, self-realization, and self-confidence by learning and experiencing new sports.

Another topic is that sports tourists try to learn about the cultural heritage of the locals in their tourist destination. Tourists can find out the host community's eating and drinking habits, wedding-death ceremonies, customs, unchanging and unspoiled habits, lifestyles, historical structure, appearances (Saçılık & Toptaş, 2017) and will curiously try to learn their cultural heritage. They also want to communicate with local people, meet new people, learn new things, and share what they know (MNE, 2014).

Conclusion

As a result, tourism is one of the most effective activities that can respond to people's physical, social, and psychological needs directly and indirectly for a certain period of time (Usta, 2002, p. 2).

There are different motivations in tourism that inspire people to participate in different tourist activities. These motivations are resting, physical and spiritual self-renewal, moving away from the current place for a short time, visiting historical and cultural areas, being close to nature, entertainment, meeting different people, exploring adventure, and participating in sports (İçöz & Kozak, 2002, p. 89). More than one motive can lead to a behavior at the same time. A person who wants to go on a holiday may have different motivations, such as being with loved ones, resting, doing sports, and getting rid of frustration (Garda, 2010). Most of these motives can be met by doing sports tourism.

Sports tourism has positive effects both for tourists and local people. Individuals coming from different societies exchange information with each other and recognize their cultural structures more closely and make friendships. As a result of these established friendships, it will be possible to contribute to world peace. The desire to communicate with strangers will encourage research and learning languages. People will enrich their own culture with different cultures (Civelek, 2010).

The fact that the result of a sports tourism activity is unclear increases the enthusiasm for participation. Uncertainty initiates a struggle, but it also creates concern and a sense of risk. This situation may excite an individual and frighten another. The struggle can be mental, spiritual, emotional, or physical. If the nature of the activity exceeds the tourist's ability, the experience can turn into misfortune or tragedy. If the tourist sees the experience as an adventure, they feel optimistic and think that success is possible. If the tourist sets a target and cannot achieve it, they can still hold the experience as a good memory (Garda, 2010).

All aspects of sports tourism will be of interest as long as they are new and untested, or if the previous experiences have been improved. Innovation contributes to the sense of escape from reality. If an experience is simple and predictable, it won't be an adventure. Experiences involving innovation awakes more emotions, such as feelings, excitement, enthusiasm, fear, and anxiety. Tourists who experienced high emotions will revitalize others and raise their awareness. The difficulty in some parts of the sports event leads to emotional confusion in most tourists, such as horror and joy, happiness and sorrow, anxiety and pleasure (Garda, 2010). Additionally, the challenge is an element that can be used to distinguish sports tourism (Cater, 2005, p. 322).

Therefore, it can be said that sports tourism has many benefits in terms of social, physical, mental well-being, and education. People who participate in sports tourism have many opportunities to socialize, to communicate, to obtain physical strength, to prove themself, to push their boundaries, to get excited, to self-realize, to satisfy one's curiosity, to develop positive mental attitudes, to learn new things, to recognize new cultures, and to make new friendships. In this context, sports tourism should not only be considered as traveling from one place to another, watching or participation, but it should also be taken into consideration that human beings have various opportunities and benefits.

References

Akman, A. D. (2007). *Turizm gelişmesinin yarattığı doğal ve kültürel değişimler: Kaş örneği [Natural and cultural changes created by tourism development: Kas example].* [Unpublished master's thesis] Ankara University Social Sciences Institute, Ankara.

Ap, J., & Crompton, J. L. (1993). Resident's strategies for responding to tourism impacts. *Journal of Travel Research, 32*(1), 47–50.

Argan, M. (2004). Spor ve turizm pazarlamasının kesişim noktası olarak spor turizmine kuramsal bir bakış [A theoretical approach to sports tourism as a point of intersection of sports and tourism marketing]. *Anatolia: Journal of Tourism Research, 15*(2), 158–168.

Batat, W., & Frochot, I. (2014). Towards an experiential approach in tourism studies. In S. McCabe (Ed.), *The Routledge handbook of tourism marketing* (pp. 109–123). Routledge.

Beedie, P., & Hudson, S. (2003). Emergence of mountain-based adventure tourism. *Annals of Tourism Research, 30*(3), 625–643.

Bektaş, F. (2010). *Kaçkar havzası trekking parkurlarının spor turizmi bakımından değerlendirilmesi [Evaluation of trekking trails in terms of sports tourism in Kackar catchment].* [Unpublished doctoral dissertation] Gazi University Education Sciences Institute, Ankara.

Brunt, P., & Courtney, P. (1999). Host perceptions of socio-cultural impacts. *Annals of Tourism Research, 26*, 493–515.

Cater, C. (2005). Looking the part: The relation between adventure tourism and the outdoor fashion industry. In M. Aicken (Ed.), *Taking tourism to the limits: Issue, concepts and managerial perspectives* (1st ed., pp. 155–165). Elsevier Science Ltd.

Choi, S. (1999). Benefiting from mega-events: Olympics 2000, World Cup 2002 and 2005 World Exposition. *Journal of Sport Tourism, 5*(4), 29–35.

Civelek, A. (2010). Turizmin sosyal yapıya ve sosyal değişmeye etkileri [The effects of tourism on social structure and social change]. *Journal of Selçuk University Social Sciences MYO, 13*(1–2), 331–350.

Crompton, J. (1979). Motivations for pleasure vacation. *Annals of Tourism Research, 6*, 408–424.

Deccio, C., & Baloglu, S. (2002). Nonhost community resident reactions to the 2002 Winter Olympics: The spillover impacts. *Journal of Travel Research, 41*(1), 46–56.

Delamere, T. A. (2001). Development of a scale to measure resident attitudes toward the social impacts of community festivals, part II: Verification of the scale. *Event Management, 7*(1), 25–38. https://doi.org/10.3727/152599501108751452

Demircioğlu, A. G. (1997). Turizm çevre etkileşimi bakımından sürdürülebilir turizm planlaması [Sustainable tourism planning in terms of tourism environment interaction]. *Journal of Dokuz Eylül University İİBF, 12*(2), 135–147.

Fagence, M. (2003). *Tourism in destination communities.* Cabi Publishing.

Fredline, E. (2005). Host and guest relations and sport tourism. *Sport, Culture and Society, 8*(2), 263–279.

Garda, B. (2010). *Macera turizmi pazarlaması: Antalya yöresine gelen turistlerin macera turizmine yönelik eğilimleri üzerine bir araştırma [Adventure tourism marketing: A research on the tendency of tourists coming to Antalya region towards adventure tourism].* [Unpublished doctoral dissertation] Selcuk University, Konya.

Garda, B. (2014). *Macera turizmi pazarlaması özel ilgi turizminin yeni yüzü [Adventure tourism marketing is the new face of special interest tourism].* Çizgi Bookstore Publishers.

Garda, B., & Temizel, M. (2016). Sürdürülebilir turizm çeşitleri [Types of sustainable tourism]. *Journal of Selcuk University Social and Technical Researches, 12,* 83–103.

Gibson, H. (1998). Sport tourism: A critical analysis of research. *Sport Management Review, 1,* 45–76.

Goeldner, C. R., & Long, P. T. (1987). The role and impact of mega-events and attractions on tourism development in North America. *Proceedings of the 37th congress of AIEST, 28.*

Goldsmith, R. E., & Tsiotsou, R. H. (2012). Introduction to experiential marketing. In R. H. Tsiotsou & R. E. Goldsmith (Eds.), *Strategic marketing in tourism services* (pp. 207–214). Emerald Group Publishing Limited.

Griffin, K., & Stacey, J. (2011). Towards a "tourism for all" policy for Ireland: Achieving real sustainability in Irish tourism. *Current Issues in Tourism, 14*(5), 431–444.

Gürsoy, Y. (2016). Giresun ilinin spor turizmi açısından değerlendirilmesi [Evaluation of Giresun province in terms of sports tourism]. *The Journal of International Social Research, 9*(45), 1080–1088.

Hall, S., Marciani, L., & Cooper, W. E. (2008). Emergency management and planning at major sports events. *Journal of Emergency Management, 6*(1), 6–15.

Hartmann, K. D. (1982). *Zur psychologie des Landschaftserlebens.* Studienkreis für Tourismus: Starnberg.

Higham, J. (1999). Sport as an avenue of tourism development: An analysis of the positive and negative impacts of sport tourism. *Current Issues in Tourism, 2*(1), 82–90.

İçöz, O., & Kozak, N. (2002). *Turizm ekonomisi [Tourism economy].* Turhan Publisher.

Iso-Ahola, S. E. (1982). Towards a social psychological theory of tourism motivation: A rejoinder. *Annals of Tourism Research, 9*(2), 256–262. https://doi.org/10.1016/0160-7383(82)90049-4

Kasap, A. A., & Faiz, G. (2008). *Bir endüstri olarak golf [Golf as an industry].* Publishers of Turkish Golf Federation.

Kim, W. (2012). *Development of a scale to measure local residents' perceived social impacts of hosting large-scale sport events* [Unpublished doctoral dissertation]. The University of Southern Mississippi, Hattiesburg, MS.

Lankford, S. V., & Howard, D. R. (1994). Developing a tourism impact attitude scale. *Annals of Tourism Research, 21*(1), 121–139.

MacCannell, D. (1977). *The tourist.* Schockon.

Markwell, K. (2004). Constructing, presenting and interpreting nature: A case study of a nature based tour to Borneo. *Annals of Leisure Research, 7*(1), 19–33.

McGee, N. G., Loker-Murphy, L., & Uysal, M. (1996). The Australian international pleasure travel market: Motivations from a gendered perspective. *The Journal of Tourism Studies, 7*(1), 45–47.

Ministry of National Education [MNE]. (2014). *Eğlence hizmetleri turizm ve insan psikolojisi [Entertainment services of tourism and human psychology].* MNE Publishers.

Nogowa, H., Yamguchi, Y., & Hagi, Y. (1996). An empirical research study on Japanese sport tourism in sport-for-all events: Case studies of a single night event and a multiple-night event. *Journal of Travel Research, 35*(2), 46–54.

Pearce, G. D. (1989). *Tourism development* (2nd ed.). Longman.

Pizam, A., & Milman, A. (1986). The social impacts of tourism. *Tourism Recreation Research, 11*(1), 29–33.

Pugh, C., & Wood, E. (2004). The strategic use of events within local government: A study of London Borough Councils. *Event Management, 9*(1/2), 61–71.

Ratz, T. (1997). *The socio-cultural impacts of tourism* [PhD Research Project]. Budapest University Economic Survey. Retrieved from http://www.ratztamara.com/impacts.html

Ritchie, B. W., Shipway, R., & Cleeve, B. (2009). Resident perceptions of mega sporting events: A non-host city perspective of the 2012 London Olympic Games. *Journal of Sport and Tourism, 14*, 143–167.

Rızaoğlu, B. (2004). *Turizm ve toplumsallaşma [Tourism and socialization]* (3rd ed.). Detay Publisher.

Ross, D. S. (2001). *Developing sports tourism.* National Laboratory for Tourism and eCommerce.

Saçılık, Y. M., & Toptaş, A. (2017). Kültür turizmi ve etkileri konusunda turizm öğrencilerinin algılarının belirlenmesi [Determination of tourism students' perceptions about cultural tourism and its effects]. *Journal of Tourism Academic, 4*(2), 107–119.

Sarı, Y., & Özdemir, C. (2014). Turizm Gelişiminin Eskişehir Odunpazarı Sakinlerinin Yaşam Kalitesi Üzerindeki Etkisi [The impact of tourism development on the quality life of Eskişehir Odunpazarı residents]. *Proceedings of 15. The National Tourism Congress,* Ankara.

Schmidhauser, H. (1989). Tourist needs and motivations. In S. F. Witt & L. Moutinho (Eds.), *Tourism marketing and management handbook* (pp. 569–572). Prentice Hall.

Sharpley, R. (1994). *Tourism, tourists and society.* ELM Publisher.

Standeven, J., & de Knop, P. (1999). *Sport tourism.* Human Kinetics.

Taylor, T. L., & Toohey, K. M. (2007). Perceptions of terrorism threats at the 2004 Olympic Games: Implications for sport events. *Journal of Sport Tourism, 12*(2), 99–114.

Travis, S. A. (1984). Social and cultural aspects of tourism. *UNEP Industry and Environment, 7*(1), 22–24.

Usta, Ö. (2002). *Genel turizm [General tourism].* Anadolu Publisher.

15 Physiological Effects of Active Tourism

Petra Mayer, Zoltán Ádám, and Marta Wilhelm

Impact of Climate Change on Tourism

Tourism can serve various socio-cultural, economic, and environmental purposes. People around the world travel for sports, recreation, religious practices, medical-health activities, and many other reasons. It is well known that active travel provides an extensive range of health benefits, both physically and mentally.

Vacation itself has positive effects on health and well-being, but these effects can soon fade out after resumption to work (De Bloom et al., 2009). Three days after vacation, physical complaints, quality of sleep, and mood had improved as compared to before vacation, and five weeks after holiday vacationers still reported fewer physical complaints than before (Strauss-Blasche et al., 2000). The duration of vacation can be determinative, but according to De Bloom et al. (2013), in vacations longer than 14 days, health and well-being increased quickly during vacation, peaked on the eighth vacation day, and had rapidly returned to baseline level within the first week of work resumption. They found that vacation duration and most vacation activities were only weakly associated with health and well-being changes during and after vacation. Teachers before and after vacation reported that the positive effects of vacation on work engagement and burn out faded out within one month. Job demands after vacation sped up this process (Kühnel & Sonnentag, 2011). To have a real recovery, we need a lot of free time for ourselves, choosing a vacation location with more sun. Exercises during vacation, good sleep, and making new acquaintances promote recuperation. Exhaustion is increased by vacation-related health problems and a greater time-zone difference to home, but reduced with warmer vacation locations (Strauss-Blasche et al., 2005).

Short-term vacations also have an impact on health if there are various physical activity (PA) programs within. Neumayr and Lechleitner (2018) studied the effects of low to moderate intensity exercises on cardiovascular parameters before and after a one-week-long vacation on healthy subjects. Participants performed two types of vacation activities: in the first group, subjects played golf for 33.5 h/week, and in the second group volunteers engaged in Nordic walking or e-biking for 14.2 h/week. Cardiovascular parameters such as performance capacity, blood pressure (BP), heart rate (HR) profiles and cardiac diastolic function (DF) were measured. There was a significant decrease in body weight of 1.0 kg in the Nordic walking

DOI: 10.4324/9781003476658-19

and e-bike group but not in the golf group, while they noted a reduction of BP and HR in both groups, which was significant only in the golf program. In addition, an improvement of cardiac DF was measured in both groups; with a more pronounced improvement in the Nordic walking and e-bike group. The benefits were probably due to the enhanced PA rather than purely holiday effects (Neumayr et al., 2014). Moderate-intensity activities are feasible for nearly everybody, including persons with poor cardiorespiratory fitness. A nine-month-long mountain hiking program was completed by elderly participants, consisting of a single weekly hiking session with the goal of achieving a 500-m altitude increase within 3 h (Gatterer et al., 2015). Before and after the nine-month program, an electrocardiogram was performed and BP, glycated hemoglobin, high-density (HDL) and low-density lipoprotein (LDL) measurements were also done. No changes were found in any of the investigated parameters for the entire group; however, participants with untreated hypertension showed a reduced systolic BP. The authors noted that moderate-intensity activity only once a week does not improve cardiovascular risk factors in elderly persons with a relatively normal cardiovascular risk profile. Conversely, elderly persons suffering from hypertension might profit from such a practice.

In recent years, the popularity of forests themselves as a recreational place (hiking and tracking) has gained attention (Dudek, 2017). Most of the time people from urban areas select these destinations due to their needs and interest toward nature. Several studies reported that these destinations work as a therapy for stress release and are beneficial for human health (Park et al., 2009; Karjalainen et al., 2010). Bird watching, hiking are some popular activities in these areas.

Walking and hiking is a popular recreational and tourism activity in many countries with a lot of benefits on health. Not only the duration of the trip is determinative, but altitude itself also has many physiological effects.

Thermal comfort indices have been developed in order to capture the complexity of the thermal aspect of climate, which is argued to be a composite of temperature, wind, humidity, and radiation. In a case study, De Freitas (2003) found that the optimal thermal conditions for beach users are not at the minimum heat stress level but at a point of mild heat stress. Using the thermal comfort index, Thorsson et al. (2004) found a positive relationship between thermal comfort and urban park use of recreational activities in Göteborg, Sweden.

Hu and Ritchie (1993) reviewed several studies and finding in their own study (examined Hawaii, Australia, Greece, France, and China using a survey of Canadian citizens) that climate is the second most important characteristic for the group of tourists in a recreational holiday.

Altitude Studies

In mountain regions, climate change can have severe impacts. For example, the rise in temperature of Himalayan regions is recorded in the range from 0.06° C per year to 0.27° C per year (Shrestha et al., 2012). This rise in temperature has several implications, such as the melting of glaciers, increased pests, changes in precipitation, and shifting seasons.

However, more and more people decide on vacations with hiking, long-distance walking, and running. In a study, male patients with metabolic syndrome were participating in a three-week vacation program. Halve of them in a moderate altitude group (at 1,700 m), and the others in a low altitude group (at 200 m). The program included 12 hiking tours (four per week, average duration 2.5 hours, intensity 55–65% of HR maximum). Physical parameters, performance capacity, 24-hour BP, and HR profiles were obtained before, during, and after the stay. A significant mean weight loss of 3.13 kg was found in both groups, but changes in performance capacity were minor. Systolic, diastolic, and mean arterial pressures and circadian HR profiles were significantly reduced in both groups, with no differences between them. Consequently, the pressure-rate product was reduced as well. All study participants tolerated the vacation well without any adverse events (Neumayr et al., 2014). In the study of Greie et al. (2006), the same results were obtained in a very similar investigation, but in the low altitude group, fasting insulin and homeostasis model assessment (HOMA)-index were significantly decreased one week after return (Greie et al., 2006). During adaptation to moderate altitude, persons with metabolic syndrome exhibit an increase in erythropoiesis and a rightward shift of the oxygen dissociation curve that is similar to healthy subjects (Schobersberger et al., 2005). Researchers suggested that a two-week hiking vacation at moderate altitude may be more beneficial for adipokines and parameters of lipid metabolism than training at low altitude (Gutwenger et al., 2015).

Long-distance hiking trails, with moderate intensity, are recommended for healthy people. The duration of these hiking trips might be more than five weeks, sometimes lasting many months. In two case studies, researchers studied the alteration of the physiological parameters during more than five-week-long hiking. Devoe et al. (2009) evaluated the effects of a long-distance backpacking trip on body composition, weight, blood lipids, and lipoproteins of a 49-year-old experienced backpacker who hiked 118 days on at the Appalachian Trail. They found that an extended backpacking adventure can noticeably reduce and clinically normalize blood lipids and lipoproteins without medication and can very positively affect body composition and weight (Devoe et al., 2009). After a 34 days hiking on the Colorado Trail, from initial to post-hike testing, the body weight of participants decreased 5%, body fat decreased 1.2%, total cholesterol increased (10%), while triglycerides and HDL decreased (29%, 2%), LDL increased (23%). Resting HR decreased from 85 bpm to 67 bpm (21%). The systolic (27%) and diastolic BP decreased (3%), while VO_{2max} increased (17%) (Paradise, 2016).

Does PA during vacation have long-lasting effects on our everyday mobility behavior? Seven percent of tourists participating in research stated that they had changed their daily mobility behavior after their return, with most of them walking or cycling more often (Schlemmer et al., 2019). Another study investigated the changed PA behavior of long-distance backpackers after more than three-week-long walking trip. The authors asked backpackers of PA changes after their first long-distance hiking tour and found significant increases in the amount of everyday PA activity compared to earlier mobility behavior (Figure 15.1; Mayer & Vass, 2019).

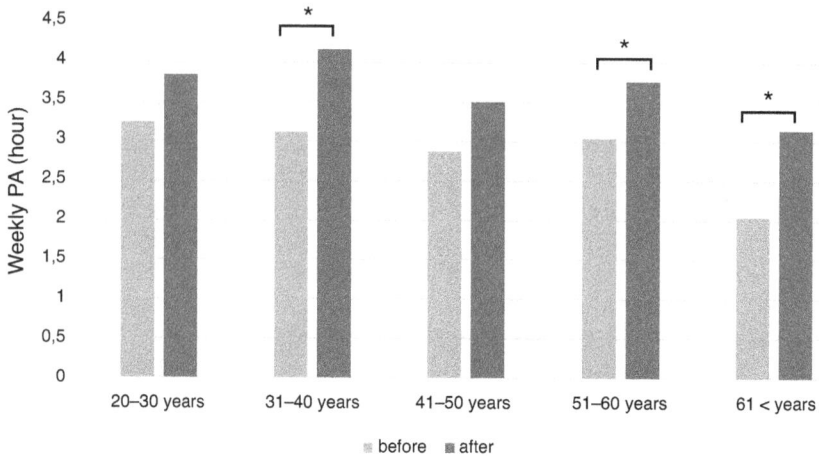

Figure 15.1 Effects of Long-Distance Hiking on Physical Activity Habits by Age Groups.

Human-Induced Climate Change: (Cambridge University Press)

Whether in the process of deciding on the destination and the right time for a holiday or in the daily choices made about recreational activities whilst on holiday, climate and weather play an important role (Hamilton & Lau, 2005). The importance of climate for tourism has been classified by de Freitas (2003) according to its aesthetic, physical, and thermal aspects. There is growing evidence, however, that climate has significant neurological and physiological effects also (Wight, 2005). Two broad types of study are in the literature where the importance of climate and weather for tourism and recreation have been examined (altitudinal studies and behavioral studies).

Disadvantages of Active Tourism

Walking, Hiking, Trekking Injuries

The popularity of recreational hiking has continuously increased since the 1940s, with millions of tourists enjoying backpacking and day hiking. Injuries are unfortunate possible consequences of hiking. It is important to know how to prevent them and to know some basic first aid when going outdoors. Multiple studies have focused on specific types of injuries hikers are facing, including load-induced musculoskeletal and metabolic injuries.

A prospective, observational cohort survey of 334 persons hiking the Appalachian Trail for more than seven days was conducted (Boulware, 2004). Medical experiences of backpackers were examined and found that the most common musculoskeletal injuries were acute joint pain (37%), back pain (25%), Achilles heel pain (23%), chronic joint pain (20%), muscle cramps (20%), muscle strain (15%), tendinitis (15%), shin splints (14%), and sprained ankle (13%). Cutaneous problems involve feet blisters (64%), paresthesia's (46%), chafing (44%), and sun-burn (39%). The occurrence of individual musculoskeletal problems and of di-arrhea (56%) were similar between sexes and there was no correlation between weight loss and diarrhea. Dehydration (20%) and hypothermia (10%) also appeared among participants. Menstrual changes were very common including amenorrhea (Boulware, 2004).

A recent systematic review focusing on hiking incidents extracted hiking injury and illness prevention recommendations and classified them. Their most useful findings were: be more aware of slippery surfaces, have basic first aid knowledge, obey warning signs, and bring the proper equipment. They found recommendations for animal attacks, for example, food should be kept in airtight storage contain-ers well away from the sleeping area, hung in or between trees, and also children should be educated about dangerous animals. According to their data, illness pre-vention advice was often related to hygiene, water disinfection, and diet. Glucose or carbohydrate replacement during prolonged exercise is recommended to avoid metabolic exhaustion and fatigue. Researchers found that group members have an important role in safety due to the determination of group norms and risk-taking behaviors (Kortenkampa et al., 2017).

Gardner and Hill (2002) asked long-distance backpackers about details of the hike, including illness and injury and their pre-hike preparation in light of their experience on the trail. There was no significant correlation noted between having a pre-training program and experiencing injury or illness during hiking. Those who did not wear or precondition their footwear before hiking had an increased risk of developing blisters (Figure 15.2). Only 6% of respondents visited a health-care practitioner before the hike to address health needs on the trail (Gardner & Hill, 2002). Most injuries and rescues occur from underestimating the risks from extrin-sic, environmental factors, and/or overestimating one's intrinsic skills. By match-ing the fitness and skill level of the hiker to the environment, physician can help to reduce the risk of serious injuries (Green, 2015).

A systematic review was conducted to determine if sock, antiperspirant, or bar-rier strategies were effective in the prevention of friction blisters in the wilderness and outdoor activities. The moderate effect of paper tape was shown as an effective form of barrier prevention (Worthing et al., 2017).

Cities and Urban Center Tourism

Advancement in technology, ease of travel through faster transportation, iconic building designs, landmarks, and modern market designs in the form of malls, provide cities distinctive tourist destinations. Contemporary building designs with planned city-style enable them to accommodate and attract a wide range of peo-ple. However, these places face climate change impacts in the form of heatwaves, smog, changing beach zones, and seasonal shifts. The buildings and road material themselves make these processes accelerate faster.

Most of the urban areas in tropical and semi-arid zones often experience a regional climate phenomenon known as the Urban Heat Island (UHI, Li et al., 2005).

Figure 15.2 Different Types of Pain Relief after Long Hikes.
Source: Gifts of Sz. Koncz (2021).

UHI is observed as the metropolitan area having significantly warmer temperature than its surrounding rural areas. In such cases, air trapped between the tall building and narrow streets can heat up (Figure 15.3). Additionally, the concrete land cover, infrastructure, and industrial activities add as catalysts to this process (Weng & Larson, 2005).

Age, Fitness, and Regional Blood Flow during Exercise in the Heat

Several factors affect the responses of an individual to combined stresses of exercise and heat including age, anthropometric characteristics, maximal aerobic capacity, and the level of acclimatization (Inbar et al., 2004). Typically, young children and older people have low maximal aerobic power, high adiposity and small body stature and body mass compared to young adults. These morphological and physiological characteristics imply relatively large surface area-to-mass (especially in children), lower sweat rate, lower cardiac output, and poor control of peripheral blood flow (typically in older individuals, Figure 15.4) compared to young healthy adults (Pandolf, 1991; Kenney, 1997).

Figure 15.3 Urban Recreation, Relief after Long City Tours.

Source: Gifts of Sz. Koncz (2020).

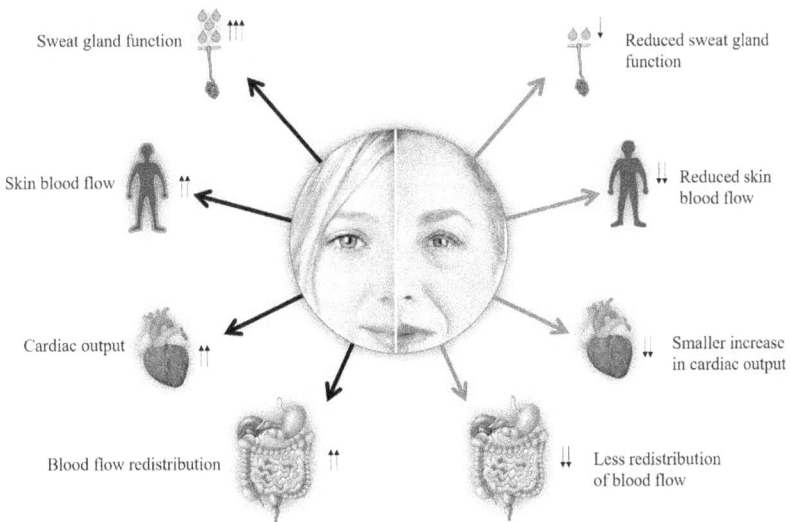

Figure 15.4 Summary of the Age-Related Changes in Thermoregulation of the Body under Heat Stress.

Source: Based on Kenney & Munce (2003).

Although most research on human thermoregulation has focused on young adult males, few studies compared the thermoregulatory responses between children and adults. Only one study investigated the age-related differences in thermoregulation during exercise involving all three lifespan stages (childhood, adulthood and old age, Drinkwater & Horvath, 1979). They compared exercise-induced thermoregulation responses between prepubertal girls and elderly women. Their findings showed that prepubertal girls are less tolerant of the strain imposed by exercise in hot environments than adult women. The reasons of this phenomena may be the circulatory instability caused by the shift in blood volume from the central to the peripheral circulation, higher skin temperature due to delayed onset of sweating, and a larger surface area relative to blood volume.

In the literature few studies suggest that thermoregulation in children is just as effective as that of adults (Davies, 1981; Docherty et al., 1986), a great majority of others indicate less effective thermoregulation in case of children (Wagner et al., 1972; Sloan & Keatinge, 1973; Drinkwater et al., 1977; Delamarche et al., 1990; Havenith, 2001).

Apparently, the most notable differences between the thermoregulation of the two age groups originate in the lower overall sweating rate in children (Delamarche et al., 1990) and in reduced peripheral blood flow (Inoue et al., 1999). These differences may contribute to the greater rate of increase in body temperature and reduced heat tolerance in young and elderly individuals.

The increasing number of children involved in certain types of high-level PA and those elderly who participate in regular exercise or recreational programs to maintain or improve their health may rise the risk of disorders related to heat in individuals of these two lifespan stages (Figure 15.5).

In case of elderly, sweat gland productivity was high, while a relatively inefficient cooling rate was observed mainly due to the low overall sweating rate (Inbar et al., 2004). Older individuals have a reduced rate of sweat secretion and delayed onset of sweating during periods of increased core temperature and heat stress (Hellon & Lind, 1956; Wagner et al., 1972). The decrease in cooling capacity has been suggested to be caused by an age-related decrease in the sensitivity of cholinergic stimulation and reduced androgenic stimulation of the secretory coil, which result in a reduced sweating capacity in the elderly population (Kenney, 1997). The lower peripheral blood flow in response to exercise caused heat stress may be due to the lower final HR (to measured peak HR) as well as the slow and low increase in HR which are characteristics of low peak cardiac output also (Inbar et al., 2004).

Effect of Forest Bathing Trips on Human Immune Function

There is a deep connection between all the living and nature. In humans, this biophyl trait materializes in an instinctive craving for nature which is becoming increasingly significant in parallel to the increasing rate of urbanization (Li, 2010).

The forest and off urban environment have been enjoyed by people for countless reasons. The beautiful scenery, the tranquil atmosphere, moderate climate, fresh air has beneficial effects physically and mentally as well (Yamaguchi et al., 2006),

Figure 15.5 Resting and Cooling Strategies of Older People Depend on Experience and Fitness Level.

while people living in an urban environment usually develop stress-related fatigue, and respiratory problems (Yamaguchi et al., 2006). In Japan, a popular short lei-surely visit to the forest or "forest bathing" has stood the test of time. These trips serve as tranquilizer and recreational purposes while participants breath in volatile substances. Those aroma therapeutics are wood essential oils, called phytonicides, having antimicrobial and immune-strengthening effects too (Li et al., 2006).

Li (2010) showed the effect of a three days/two-nights forest bathing trip on the efficiency of the human immune system. They showed significant elevation in NK cell activity which was supported by the significant increase of serum level of metabolites (intracellular perforin, granzyme, etc.) produced by this cell type. It has been reported that NK cells have a major role in eliminating tumor and virus-infected cells through the release of substances playing a role in the perforin/granzyme apoptosis pathway via granule exocytosis (Okada et al., 2003).

During certain circumstances (novelty, anticipation, unpredictability, emotional and physiological stress), the level of adrenaline released by the adrenal medulla is increasing. Under increased PA noradrenalin level, the predominant neurotransmitter of the sympathetic system increases; some of it enters the circulation resulting in an increased detectable level (Frankenhaeuser, 1975). This has been used to evaluate the stress level related to work in truck drivers (van der Beek et al., 1995) and nurses (Brown et al., 2006) from urine samples. Forest bathing trips proved to have an exceptional effect on the concentration of these hormones in both male (Li et al., 2008b) and female individuals (Li et al., 2008a), while city tours are shown to be ineffective in that matter, suggesting that subjects experience lower stress during trips in a forest environment (Li et al., 2008a). Li et al. (2008a) also reported a significant increase in the proportions of lymphocytes and monocytes, while the proportions of granulocytes decreased in peripheral blood, which might be due to an increase in parasympathetic dominance previously reported by Mori et al. (2002).

Many reports state that a forest environment has exceptional effects on patients suffering from allergies or respiratory diseases (Loureiro et al., 2005).

Programs for Elderly

Canadian statistics estimate that by the year 2030, the ratio between young and elderly individuals in Canada will change in favor of the elderly and similar estimations exist in most modern societies all over the world. Among the elder generation, fear of falling is ranked highest among cases like financial problems and criminal violence (McKinley et al., 2008). This fear hinders their outdoor activities and efforts to socialize (McAuley et al., 2000). Moreover, this state of anxiety decreases satisfaction with their life and increases disability and enhances isolation from community-dwelling (McAuley et al., 2000).

McKinley et al. (2008) showed in their feasibility study that Argentine tango dance classes compared to walking sessions as PA is feasible in the case of the elderly population to increase their level of community-dwelling, physical fitness, and balance.

Fisher and Li (2004) created a project called Senior Health and Physical Exercise (SHAPE) in which they aimed to evaluate the effect of a neighborhood-level walking intervention. They found that the six-month study improved their quality of life indicators (physical functioning, mental well-being, and physical functioning) compared to inactive controls.

A recent study found that approximately one-third of cardiac diseases and osteoporosis cases could be related to lack of PA (Garrett et al., 2004). There is a relationship between low cardiorespiratory fitness, poor functional performance (Binder et al., 1999), and increased risk of cardiovascular diseases (Kurl et al., 2003). Fitness and exercise programs proved to be effective to improve the quality of life of the elderly even after cardiovascular sicknesses, like stroke (Pang et al., 2005).

Clinical Effects of Regular Dry Sauna Bathing

During exercise, body temperature will rise due to the metabolic rate increase of muscles. Blood circulating through the muscles becomes warmer, resulting in an increased core temperature. The amount of heat produced is related to the amount of work performed by the muscles. The more vigorous the workout, the more heat is produced.

Sauna bathing as a form of whole-body thermotherapy has been used in various forms (sweat lodges, radiant heat, etc.) for a long time for several reasons (hygiene, health, spiritual, and social gathering) (Hussain & Cohen, 2018). The relationship between sauna usage and exercise which may have synergistic hormetic responses is still an area of active research (Hussain & Cohen, 2018); however, it seems beneficial in the long run. There is considerable evidence suggesting that sauna bathing can induce numerous physiological effects (Hussain & Cohen, 2018). The short-term intensive heat exposure increases the peripheral and core temperature and activates hypothalamic thermoregulatory pathways (Zhao et al., 2017) and increases the sympathetic modulation through the central nervous system (CNS). Through this activation (hypothalamus-pituitary-adrenal hormonal axis, and the renin-angiotensin-aldosterone axis), the sympathetic tone increasing, resulting in increased HR, peripheral and skin blood flow (reduced vascular resistance), sweating and cardiac output (decreased systolic/diastolic BP) (Hussain & Cohen, 2018). On a cellular basis, heat therapy induces metabolic changes as well (increased production of HSP, insulin sensitivity, reduction of ROS, oxidative stress, inflammation, etc., Iguchi et al., 2012). It has been suggested that there is a similarity between the mechanisms of exercise and heat-induced adaptive hormesis; however, these mechanisms are still being explored (Hussain & Cohen, 2018). Apart from the physiological benefits, sauna bathing has been reported to have positive psychological impacts (maybe due to the release of endorphins and other opioid-like peptides) (Hussain & Cohen, 2018), like stress reduction, relaxation, forced mindfulness, and improved sleep just to name a few. Sauna has been shown to reduce cholesterol and LDL levels in healthy men (Gryka et al., 2014) and women (Pilch et al., 2014) after four weeks of regular sauna bathing.

References

Binder, E. F., Birge, S. J., Spina, R., Ehsani, A. A., Brown, M., Sinacore, D. R., & Kohrt, W. M. (1999). Peak aerobic power is an important component of physical performance in older women. *Journals of Gerontology Series A: Biomedical Sciences and Medical Sciences, 54*(7), 353–356.

Boulware, D. R. (2004). Gender differences among long-distance backpackers: A prospective study of women Appalachian Trail backpackers. *Wilderness & Environmental Medicine, 15,* 175–180.

Brown, D. E., James, G. D., & Mills, P. S. (2006). Occupational differences in job strain and physiological stress: Female nurses and school teachers in Hawaii. *Psychosomatic Medicine, 68,* 524–530.

Davies, C. T. M. (1981). Thermal responses to exercise in children. *Ergonomics, 24,* 55–61.

De Bloom, J., Kompier, M., Geurts, S., De Weerth, C., Taris, T., & Sonnentag, S. (2009). Do we recover from vacation? Meta-analysis of vacation effects on health and well-being. *Journal of Occupational Health, 51*(1), 13–25.

De Bloom, J., Geurts, S. E., & Kompier, M. J. (2013). Vacation (after-) effects on employee health and well-being, and the role of vacation activities, experiences and sleep. *Journal of Happiness Studies, 14*(2), 613–633.

De Freitas, C. R. (2003). Tourism climatology: Evaluating environmental information for decision making and business planning in the recreation and tourism sector. *International Journal of Biometeorology, 48,* 45–54.

Delamarche, P., Bittel, J., Lacour, J. R., & Flandrois, R. (1990). Thermoregulation at rest and during exercise in prepubertal boys. *European Journal of Applied Physiology, 60,* 436–440.

Devoe, D., Israel, R. G., Lipsey, T., & Voyles, W. (2009). A long-duration (118-day) backpacking trip (2669 km) normalizes lipids without medication: A case study. *Wilderness & Environmental Medicine, 20,* 347–352.

Docherty, D., Eckerson, J. D., & Hayward, J. S. (1986). Physique and thermoregulation in prepubertal males during exercise in warm, humid environment. *American Journal of Physical Anthropology, 70,* 19–23.

Drinkwater, B. A., & Horvath, S. M. (1979). Heat tolerance and aging. *Medicine & Science in Sports Exercise, 11,* 49–55.

Drinkwater, B. L., Kupprat, I. C., Denton, I. E., Christ, I. L., & Horvath, S. M. (1977). Response of prepubertal girls and college women to work in the heat. *Journal of Applied Physiology, 43,* 1046–1053.

Dudek, T. (2017). Recreational potential as an indicator of accessibility control in protected mountain forest areas. *Journal of Mountain Science, 14*(7), 1419–1427.

Fisher, K. J., Li, F., Michael, Y., & Cleveland, M. (2004). Neighborhood-level influences on physical activity among older adults: a multilevel analysis. *Journal of Aging and Physical Activity, 12*(1), 45–63.

Frankenhaeuser, M. (1975). Experimental approach to the study of catecholamines and emotion. In L. Levi (Ed.), *Emotions, their parameters and measurement* (p. 209). Raven Press.

Gardner, T. B, & Hill, D. R. (2002). Illness and injury among long distance hikers on the Long Trail, Vermont. *Wilderness & Environmental Medicine, 13,* 131–134.

Garrett, N. A., Brasure, M., & Schmitz, K. H. (2004). Physical inactivity. Direct cost to a health plan. *American Journal of Preventative Medicine, 27,* 304–309.

Gatterer, H., Raab, C., Pramsohler, S., Faulhaber, M., Burtscher, M., & Netzer, N. (2015). Effect of weekly hiking on cardiovascular risk factors in the elderly. *Zeitschrift für Gerontologie und Geriatrie, 48,* 150–153.

Green, G. A. (2015). Setting, structure, and timing of the preparticipation examination: The wilderness adventure consultation. *Wilderness & Environmental Medicine, 26,* 4–9.

Greie, S., Humpeler, E., & Gunga, H. C. (2006). Improvement of metabolic syndrome markers through altitude specific hiking vacation. *Journal of Endocrinological Investigation, 29,* 497–504.

Gryka, D., Pilch, W., Szarek, M., Szygula, Z., & Tota, Ł. (2014). The effect of sauna bathing on lipid profile in young, physically active, male subjects. *International Journal of Occupational Medicine and Environmental Health, 27*(4), 608–618. https://doi.org/10.2478/s13382-014-0281-9

Gutwenger, I., Hofer, G., Gutwenger, A. K., Sandri, M., & Wiedermann, C. J. (2015). Pilot study on the effects of a 2-week hiking vacation at moderate versus low altitude on plasma parameters of carbohydrate and lipid metabolism in patients with metabolic syndrome. *BMC Research Notes, 8,* 103.

Hamilton, J. M., & Lau, M. (2005). The role of climate information in tourist destination choice decision-making. In S. Gossling & C. M. Hall (Eds.), *Tourism and global environmental change* (pp. 229–250). Routledge.

Havenith, G. (2001). Human surface to mass ratio and body core temperature in exercise heat stress – A concept revisited. *Journal of Thermal Biology, 26,* 387–393.

Hellon, R. F., & Lind, A. R. (1956). Observations on the activity of sweat glands with special reference to the influence of ageing. *Journal of Physiology, 133*, 132–144.

Hu, Y., & Ritchie, J. R. B. (1993). Measuring destination attractiveness: A context approach. *Journal of Travel Research, 32*(2), 25–36.

Hussain, J., & Cohen, M. (2018). Clinical effects of regular dry sauna bathing: A systematic review. *Evidence-Based Complementary and Alternative Medicine, 2018*. https://doi.org/10.1155/2018/1857413

Iguchi, M., Littmann, A. E., Chang, S. H., Wester, L. A., Knipper, J. S., & Shields, R. K. (2012). Heat stress and cardiovascular, hormonal, and heat shock proteins in humans. *Journal of Athletic Training, 47*(2), 184–190.

Inbar, O., Morris, N., Epstein, Y., & Gass, G. (2004). Comparison of thermoregulatory responses to exercise in dry heat among prepubertal boys, young adults and older males. *Experimental Physiology, 89*(6), 691–700. https://doi.org/10.1113/expphysiol.2004.027979

Inoue, Y., Havenith, G., Kenney, W. L., Loomis, J. L., & Buskirk, E. R. (1999). Exercise and methacholine-induced sweating responses in older and younger men: Effect of heat acclimation and aerobic fitness. *International Journal of Biometeorology, 42*, 210–216.

Karjalainen, E., Sarjala, T., & Raitio, H. (2010). Promoting human health through forests: Overview and major challenges. *Environmental Health and Preventive Medicine, 15*(1), 1–8.

Kenney, W. L. (1997). Thermoregulation at rest and during exercise in healthy older adults. *Exercise Sport Sciences Review, 25*, 41–76.

Kenney, W. L., & Munce, T. A. (2003). Invited review: aging and human temperature regulation. *Journal of Applied Physiology, 95*(6), 2598–2603.

Kortenkampa, K. V., Mooreb, C. F., Sheridana, D. P., & Ahrensa, E. S. (2017). No hiking beyond this point! Hiking risk prevention recommendations in peer-reviewed literature. *Journal of Outdoor Recreation and Tourism, 20*, 67–76.

Kühnel, J., & Sonnentag, S. (2011). How long do you benefit from vacation? A closer look at the fade-out of vacation effects. *Journal of Organizational Behavior, 32*, 125–143.

Kurl, S., Laukanen, J. A., & Rauramaa, R. (2003). Cardiorespiratory fitness and the risk for stroke in men. *Archives of Internal Medicine, 163*, 1682–1688.

Li, Q. (2010). Effect of forest bathing trips on human immune function. *Environmental Health and Preventive Medicine, 15*, 9–17.

Li, Q., Nakadai, A., Matsushima, H., Miyazaki, Y., Krensky, A. M., & Kawada, T. (2006). Phytoncides (wood essential oils) induce human natural killer cell activity. *Immunopharmacology & Immunotoxicology, 28*, 319–333.

Li, W., Wang, Y., Peng, J., & Li, G. (2005). Landscape spatial changes associated with rapid urbanization in Shenzhen, China. *International Journal of Sustainable Development and World Ecology, 12*(3), 314–325.

Li, Q., Morimoto, K., Kobayashi, M., Inagaki, H., Katsumata, M., & Hirata, Y. (2008a). A forest bathing trip increases human natural killer activity and expression of anti-cancer proteins in female subjects. *Journal of Biological Regulators & Homeostatic Agents, 22*, 45–55.

Li, Q., Morimoto, K., Kobayashi, M., Inagaki, H., Katsumata, M., & Hirata, Y. (2008b). Visiting a forest, but not a city, increases human natural killer activity and expression of anti-cancer proteins. *International Journal of Immunopathology & Pharmacology, 21*, 117–128.

Loureiro, G., Rabaca, M. A., Blanco, B., Andrade, S., Chieira, C., & Pereira, C. (2005). Urban versus rural environment – Any differences in aeroallergens sensitization in an allergic population of Cova da Beira, Portugal? *Allergy & Immunology (Paris), 37*, 187–193.

Mayer, P., & Vass, L. (2019). Hosszútávú, gyalogos zarándoklatok testmozgást befolyásoló hatásai. *Magyar Sporttudományi Szemle, 2*(63), 22–28.

McAuley, E., Blissmer, B., Marquez, D. X., Jerome, G. H., Kramer, A. F., & Katula, J. (2000). Social relations, physical activity and well-being in older adults. *Preventive Medicine, 31,* 608–617.

McKinley, P. K., Jacobson, A., Leroux, A. I., Bednarczyk, V., Rossignol, M., & Fung, J. (2008). Effect of a community-based Argentine tango dance program on functional balance and confidence in older adults. *Journal of Aging and Physical Activity, 16*(4), 435–453.

Mori, H., Nishijo, K., Kawamura, H., & Abo, T. (2002). Unique immunomodulation by electro-acupuncture in humans possibly via stimulation of the autonomic nervous system. *Neuroscience Letters, 320,* 21–24.

Neumayr, G., & Lechleitner, P. (2018) Effects of a one-week vacation with various activity programs on cardiovascular parameters. *The Journal of Sports Medicine and Physical Fitness, 59*(2), 335–339.

Neumayr, G., Fries, D., Mittermayer, M., Humpeler, E., Klingler, A., Schobersberger, W., … Schmid, P. (2014). Effects of hiking at moderate and low altitude on cardiovascular parameters in male patients with metabolic syndrome: Austrian moderate altitude study. *Wilderness & Environmental Medicine, 25*(3), 329–334.

Okada, S., Li, Q., Whitin, J. C., Clayberger, C., & Krensky, A. M. (2003). Intracellular mediators of granulysin-induced cell death. *Journal of Immunology, 171,* 2556–2562.

Pandolf, K. B. (1991). Aging and heat tolerance at rest or during work. *Experimental Aging Research, 17,* 189–204.

Pang, M. Y., Eng, J. J., Dawson, A. S., McKay, H. A., & Harris, J. E. (2005). A community-based fitness and mobility exercise program for older adults with chronic stroke: A randomized, controlled trial. *Journal of the American Geriatrics Society, 53*(10), 1667–1674.

Paradise, A. (2016). *Effects of a 500 mile backpacking through hike on the performance of a competitive powerlifter: An observational case study.* [Thesis]

Park, B. J., Tsunetsugu, Y., Kasetani, T., Morikawa, T., Kagawa, T., & Miyazaki, Y. (2009). Physiological effects of forest recreation in a young conifer forest in Hinokage Town, Japan. *Silva Fennica, 43*(2), 291–301.

Pilch, W., Szyguła, Z., Tyka, A., Palka, T., Lech, G., Cison, T., & Kita, B. (2014). Effect of 30-minute sauna sessions on lipid profile in young women. *Medicina Sportiva, 18*(4), 165–171. https://doi.org/10.5604/17342260.1133107

Schlemmer, P., Blank, C., Bursa, B., Mailer, M., & Schnitzer, M. (2019). Does health-oriented tourism contribute to sustainable mobility? *Sustainability, 11*(9), 2633. https://doi.org/10.3390/su11092633

Schobersberger, W., Greie, S., & Humpeler, E. (2005). Austrian moderate altitude study (AMAS 2000): Erythropoietic activity and Hb-O2 affinity during a 3-week hiking holiday at moderate altitude in persons with metabolic syndrome. *High Altitude Medicine & Biology, 6,* 167–177.

Shrestha, U. B., Gautam, S., & Bawa, K. S. (2012). Widespread climate change in the Himalayas and associated changes in local ecosystems. *PLoS One, 7*(5), 36741.

Sloan, R. E. G., & Keatinge, W. R. (1973). Cooling rate of young people swimming in water. *Journal of Applied Physiology, 35,* 371–375.

Strauss-Blasche, G., Ekmekcioglu, C., & Marktl, W. (2000). Does vacation enable recuperation? Changes in well-being associated with time away from work. *Occupational Medicine, 50,* 167–172.

Strauss-Blasche, G., Reithofer, B., Schobersberger, W., Ekmekcioglu, C., & Marktl, W. (2005). Effect of vacation on health: Moderating factors of vacation outcome. *Journal of Travel Medicine, 12*, 94–101.

Thorsson, S., Lindqvist, M., & Lindqvist, S. (2004). Thermal bioclimatic conditions and patterns of behaviour in an urban park in Göteborg, Sweden. *International Journal of Biometeorology, 48*, 149–156. https://doi.org/10.1007/s00484-003-0189-8

Van der Beek, A. J., Meijman, T. F., Frings-Dresen, M. H., Kuiper, J. I., & Kuiper, S. (1995). Lorry drivers' work stress evaluated by catecholamines excreted in urine. *Occupational & Environmental Medicine, 52*, 464–469.

Wagner, J. A., Robinson, S., Tzankoff, S. P., & Marino, R. P. (1972). Heat tolerance and acclimatization to work in the heat in relation to age. *Journal of Applied Physiology, 33*, 616–622.

Weng, Q., & Larson R. C. (2005). Satellite remote sensing of urban heat islands: Current practice and prospects. In R. R. Jensen, J. D. Gatrell, & D. D. McLean (Eds.), *Geo-spatial technologies in urban environments* (pp. 91–111). Springer. https://doi.org/10.1007/b137912

Wight, J. (2005). [Review of the book *Physioeconomics: The basis for long-run economic growth,* by P. Parker]. *Review of Social Economy, 63*(1), 139–144. Retrieved from http://www.jstor.org/stable/29770296

Worthing, R. M., Percy, R. L., & Joslin, J. D. (2017). Prevention of friction blisters in outdoor pursuits: A systematic review. *Wilderness & Environmental Medicine, 28*(2), 139–149. https://doi.org/10.1016/j.wem.2017.03.007

Yamaguchi, M., Deguchi, M., & Miyazaki, Y. (2006). The effects of exercise in forest and urban environments on sympathetic nervous activity of normal young adults. *Journal of International Medical Research, 34*(2), 152–159.

Zhao, Z. D., Yang, W. Z., Gao, C., Fu, X., Zhang, W., Zhou, Q., … Shen, W. L. (2017). A hypothalamic circuit that controls body temperature. *Proceedings of the National Academy of Sciences of the United States of America, 114*(8), 2042–2047. https://doi.org/10.1073/pnas.1616255114

16 Health Benefits of Sport Touristic Activities

Miklos Banhidi, Kinga Nagy, and Gyongyver Lacza

Introduction

Several large-scale epidemiologic studies focused on a central question, *what kind of factors (location and activities) influence public health*? It is almost evident that, in everyday life, the natural, socio-economic, and cultural environment, active living, and medical care have strong influence on health (Rothmann et al., 2008; Marmot et al., 2012). These factors have become the most important human values.

It is also well known that modern society has caused health problems, such as poor nutrition, obesity, and inactivity (Seidell, 2007). For this reason, tourism providers strive to offer the healthiest environments possible and health-enhancing services, such as sport programs.

We assume that it is mostly people in good health who are looking for sport tourism products, hoping to reap the benefits of physical activity. Many scientists confirm that there is irrefutable evidence of the effectiveness of regular physical activity in the primary and secondary prevention of several chronic diseases (e.g., cardiovascular disease, diabetes, cancer, hypertension, obesity, depression, and osteoporosis) and premature death (Warburton et al., 2006).

In tourism research, one of the important epidemiologic tasks is to discover to what extent health problem can influence the way of establishing a new tourism destination. Each year 12 million people travel from an industrialized country to a developing country in the tropics or subtropics. These travelers experience a high rate of diarrhea caused by a wide variety of enteric pathogens acquired by ingestion of contaminated food or water. One or more pathogens are often responsible and found in the stool of most effected individuals (Black, 1990; Banhidi & Leber, 2011).

Tourism destinations that offer sport-related services fulfil many tourist expectations, such as adventure, challenge, fun activities, sport partners, and active rest. Still, many service providers do not know how sport programs can support their business, even though it is well known that active tourists are less complicated and visit even after high season. Sport programs can improve tourists' comfort and create a healthy environment. From an epidemiological point of view, sports contribute to one's health, such as developing motor skills, cardiovascular capacity, the neuroendocrine system, or even hormone levels related to joy and happiness.

DOI: 10.4324/9781003476658-20

Other popular destinations are the high mountains, which are extremely difficult for sports tourists. If hikers or skiers do not have experience and knowledge about the impacts of cold and lack of oxygen, it can become a dangerous activity. Consequently, in the last decade, many skiing injuries were registered, often overloading the local healthcare systems. Those injuries are a loss to the tourism industry, and guests need long and intensive care, leading to a new treatment concept (Sonderegger & Simmen, 2003).

Many social policies have been introduced in tourism development to prevent negative outcomes related to environmental challenges and the growing number of risky sports. For this reason, many tourism branches have expressed their interest to increase health support services. In Europe, tourism professionals dedicated the year of 2010 to "Tourism for health". It means that in this sector, all actors should focus on health tourism opportunities thank to their rich geothermal resources, but only few discussions happen on sport opportunities. Therefore, discussing to what extent sport tourism development can support public health is important.

Measuring Benefits of Physical Activities

In 2019, a survey was conducted in Hungary focusing on the benefits of sport tourism activities. In order to understand the role of physical activity during holiday, the following research questions were proposed:

- what do adult travelers expect from traveling,
- how active are travelers during their holiday,
- and how do travelers feel physically, mentally, and socially after they return home?

In the study, the following methods were selected:

- First, the formal results in sport science were collected and analyzed to understand the influence of physical activity on the human capacity. We tried to esteem the possible impact on individual health.
- We also analyzed tourist activities and to what extent bodily functions respond to physical performance. For this we used Polar heart rate monitors while touring in the mountains or cycling. We wanted to find out how dangerous hiking or biking is for one's cardiovascular capacity.
- Next, we made a survey to measure the tourist benefits of sport tourism. For this we used a questionnaire, completed by 2,965 adults mainly from southern Slovakia. The questionnaire was developed by the Sports Tourism Section of the Hungarian Sports Science Association. Among the respondents were 61.5% males and 58% women. Among the trials, 70% of participants were active workers, 23% were below the age 26, more than half of the participants live in a household of three to four people, and 17.1% live in families of two. Those living in private houses make up 47.5% of the participants, while 25.7% live in flats (blocks of flats).

For evaluation, descriptive statistics were calculated, primarily using spread sheets to evaluate the primary data. For our first step, a datasheet was created by coding participant responses. Afterward, contingency sheets were formed to ease the processing of data. The questions were divided into three groups: closed questions, using Likert's scale, closed questions and questions regarding personal information. The questionnaire concerns the vacation practices of those queried. During the processing of the data, we were attempting to discover whether two nominal or ordinal variables are interrelated.

Results

Health Benefits of Active Sport Tourism

In sport science studies, numerous data can be found on the benefits of physical activity. Impacts are concerned with the physiological, psychological, and neuroendocrine systems. Hall et al. (2002) examined healthy nonsmoking, active, and inactive undergraduate university students. Their responses to vigorous exercise were analyzed, and they discovered negative, unpleasant feelings after the transition to anaerobic metabolism. Differences in duration and intensity of physical exercise can cause different effects for examined subjects. Higher intensity exercise became maladaptive and unpleasant because of confusion of the homeostatic mechanism (Kilpatrick et al., 2007). It can also happen in sport tourism, when tourists have to face higher intensity when performing their planned biking routes or in the mountains.

In an earlier research, physically active and inactive people were compared by Parfitt et al. (1994) at 60 and 90% of maximum capacity workload. According to Parfitt et al., high-active individuals who reported more positive responses to high intense workload may have attenuated their distress schema, whereas low-actives have not. Analyzed subjects responded differently to workload. While some experience positive feelings, others can have negative affect and greater neuroendocrine and psychophysiological responses (Hardy et al., 1989).

The result of our former survey in the Niederalpl ski area in Austria (Banhidi, 2007; Banhidi & Leber, 2011) taught us another important issue. We have been monitoring beginner and advanced skiers with polar watches on the same skiing routes. The results of the heart responses (Figure 16.1) show significant differences between skiers, which can vary up to 30%. While they are skiing on a middle-difficult slope, advanced skiers have fun just coming down, and the beginners fight to do the same distance with a submaximal capacity. Thus, it is understandable that as beginners, the negative effects of physical activity appear faster which can cause more accidents.

These results offer useful information for providers because positive experiences with sport activities will determine the future physical activity of tourists (Williams et al., 2008). Tourists who believe in the health benefits of physical activity may motivate initial involvement (Teraslinna et al., 1969), although the feelings of enjoyment and well-being supply stronger motivation for continued participation in sport programs (Morgan et al., 1984). For example, many tourists

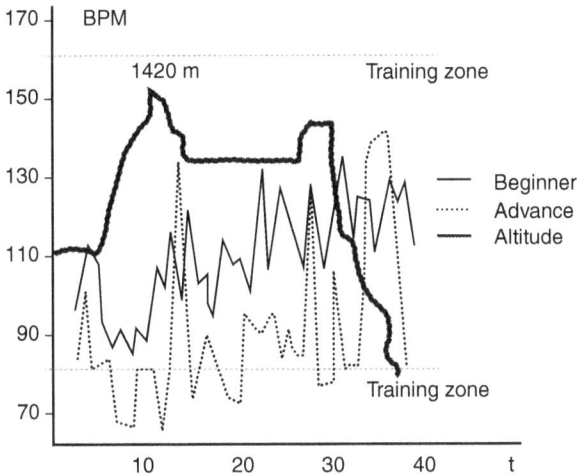

Figure 16.1 Heart Rate Responses of Beginner and Advanced Skiers at Different Altitudes in an Austrian Ski Center.

learn skiing relatively quick and experience fun and joy, which keeps them coming back to these destinations. Dishman et al. (1985) agrees that personal change may be more effective if people feel good about themselves rather than if they concentrate on knowledge of the health benefits of physical activity and exercise.

Enjoyment is an important benefit of sport tourism, which can be measured by the endorphin hormone level changes, according to Bender et al. (2007). Many forms of exercise increase the blood beta-endorphin level, principally from exercise at the anaerobic threshold and during elevated serum lactate level. Acidosis is the most effective stimulator of beta endorphin increase (Taylor et al., 1994), but only during dynamic exercise, because resistance exercise has little or no effect on beta-endorphins (Pierce et al., 1993).

Physiological responses are also present in tourists while practicing. A single bout of exercise can greatly influence insulin sensitivity (Borghouts & Keizer, 2000) but this isn't as influential as the effect of habitual physical activity. Fat, carbohydrates, lipids, glucose, and adrenalin enter the blood stream and cause stress. The stress will begin to dissipate with prolonged physical activity, but without regular exercise these bioactive compounds can damage the organism (Somogyhegyi & Nanszákné, 2006). Therefore, in accordance with previous trials, physical activity on one occasion can cause functional changes in an organism, but it is not enough for long distance health benefits.

Benefits of Holidays on Mental and Social Health Factors

Nowadays, when people are asked about their motivation for traveling and their post-holiday emotions, many of them consider new experiences as positive benefits, but at the same time, they don't like to become tired during holiday for many reasons. The focus of the survey was a holistic approach to health issues, including physical, mental, and social factors.

Usually, travelers expect to visit new countries, sights, museums, and more on their journey. According to the results, 84.5% of the respondents claimed their strongest motivation to travel was for mental rest. The explanation was that most of them want to escape from their stressful home environment.

In the survey, participants were asked how they feel when they return from holidays and if the holiday activities fulfilled their expectations. The trial had three alternative answers to choose: feeling tired, different, and refreshed or recharged.

Social health issues are always important for travelers during holidays. They feel most comfortable being together with families and friends.

The mental health factor in tourism, the experiences can be both positive and negative for residents in communities where sharing and preserving their culture could be seen as conflicting goals (Besculides et al., 2002).

The applicability of a segmentation procedure to this problem is discussed and examples are given of explanatory models of resident attitudes toward tourism's social impact. Significant differences in resident attitudes are identified and related to personal and locational characteristics, with tourist contact, length of residence, age, and language being major explanatory variables. It is argued that such differences should be given greater consideration by public and private tourism development agencies (Brougham & Butler, 1981).

As shown in Figure 16.2, 62.3% of the physically active participants feel that holidays helped them to rest mentally the most compared to other post-travel feelings categories. Nearly 48.7% of the active participants also enjoyed the physical rest, although 78.4% of them expected physically active programs during their holidays. It was a surprise for us to learn that even traveling can help tourists feel refreshed or recharged.

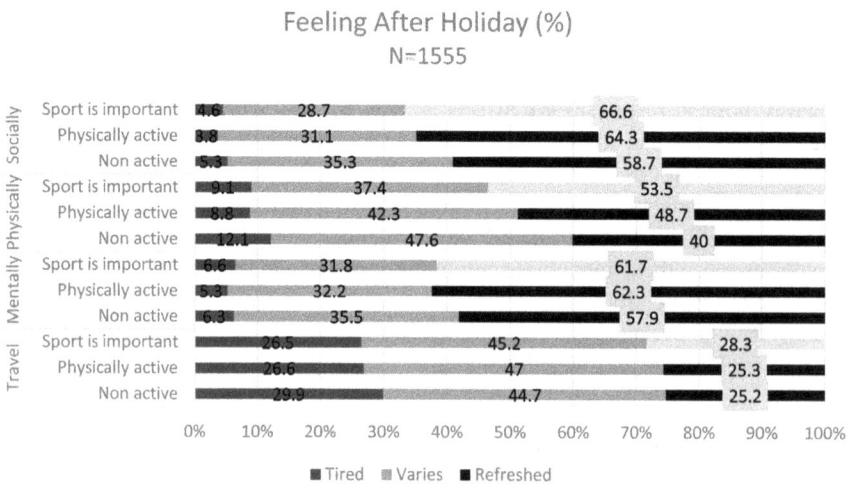

Figure 16.2 How Sport Tourists Feel about the Benefits of Their Holidays.

Conclusion

Research on sport tourism activities should be connected to health studies because it can supply additional information about the benefits of sport tourism in an extreme environment. This knowledge can contribute to reduce the number of accidents and can increase the efficiency of tourism experiences.

In this process, sport tourism research should find out more on how the environment and physical activity impacts humans. On the other hand, researchers should help find methods to prevent problems, and to help discover therapies and rehabilitation. To achieve this, there must be frequent cooperation with health tourism, because of its many benefits.

Specialists say that people who participate in regular physical activities are less sensitive to environmental changes (Banhidi & Leber, 2011); therefore, they can better manage touristic challenges. Sport tourism development is not only for human development but also to support the local destination. Even the local citizens can have the opportunity to participate in the sport tourism activities, which can offer a healthier lifestyle.

References

Banhidi, M. (2007). A sportturizmus sporttudományi modellje [Sport science model of sport tourism]. *Hungarian Sport Science Journal* [*Magyar Sporttudományi Szemle*], *30*(2), 32–38.

Banhidi, M., & Leber, R. (2011). *Sport, tourismus und umwelt in Österreich-Ungarischem kontext* [*Sport tourism and environment in Austrian-Hungarian context*]. Edtwin Wien-Győr.

Bender, T., Nagy, G., Barna, I., Tefner, I., Kádas, É., & Géher, P. (2007). The effect of physical therapy on beta-endorphin levels. *European Journal of Applied Physiology, 100*, 371–382.

Besculides, A., Lee, M. E., & McCormick, P. J. (2002). Resident's perceptions of the cultural benefits of tourism. *Annals of Tourism Research, 29*(2), 303–319.

Black, R. E. (1990). Epidemiology of travellers' diarrhea and relative importance of various pathogens. *Clinical Infectious Disease, 12*(1), 73–79.

Borghouts, L. B., & Keizer, H. A. (2000). Exercise and insulin sensitivity: A review. *International Journal of Sports Medicine, 21*(1), 1–12.

Brougham, J. E., & Butler, R. W. (1981). A segmentation analysis of resident attitudes to the social impact of tourism. *Annals of Tourism Research, 8*(4), 569–590.

Dishman, R. K., Sallis, J. F., & Orenstein, D. R. (1985). The determinants of physical activity and exercise. *Public Health Reports, 100*(2), 158–171.

Hall, E. E., Ekkekakis, P., & Petruzzello, S. J. (2002). The affective beneficence of vigorous exercise revisited. *British Journal of Health Psychology, 7*(1), 47–66.

Hardy, C. J., McMurray, R. G., & Roberts, S. (1989). AIB types and psychophysiological responses to exercise stress. *Journal of Sport & Exercise Psychology, 11*, 141–151. https://doi.org/10.1123/jsep.11.2.141

Kilpatrick, M., Kraemer, R., Bartholomew, J., Acevedo, E., & Denise, J. (2007). Affective responses to exercise are depend on intensity rather than total work. *Medicine & Science in Sports & Exercise, 39*(8), 1417–1422.

Marmot, M., Allen, J., Bell, R., Bloomer, E., & Goldblatt, P. (2012). WHO European review of social determinants of health and the health divide. *The Lancet, 380*(9846), 1011–1029.

Morgan, P. P., Shephard, R. J., & Finucane, R. (1984). Health beliefs and exercise habits in an employee fitness programme. *Canadian Journal of Applied Sport Sciences, 9*(2), 87–93.

Parfitt, G., Markland, D., & Holmes, C. (1994). Responses to physical exertion in active and inactive males and females. *Journal of Sport & Exercise Psychology, 16*(2), 178–186.

Pierce, E. F., Eastman, N. W., Tripathi, H. T., Olson, K. G., & Dewey, W. L. (1993). Plasma beta-endorphin immunoreactivity: Response to resistance exercise. *British Journal of Sports Medicine, 11*(6), 499–452.

Rothmann, K. J., Greenland, S., & Lash, T. (2008). *Modern epidemiology*. Lippincott, Williams & Wilkins.

Seidell, J. C. (2007). Epidemiology – Definition and classification of obesity. In P. G. Kopelman, I. D. Caterson, & W. H. Dietz (Eds.), *Clinical obesity in adults and children* (2nd ed., pp. 3–11). Blackwell Publishing.

Somogyhegyi, A., & Nanszákné, C. I. (2006). Connetion of inactive lifestyle with diseases. In K. Barabás (Ed.), *Egészségfejlesztés – Alapismeretek pedagógusok számára* (p. 134). Medicina Publisher.

Sonderegger, J., & Simmen, H. P. (2003). Epidemiology, treatment and results of proximal humeral fractures: Experience of a district hospital in a sports- and tourism area. *Zentralbl Chir, 128*(2), 119–124.

Taylor, D. V., Boyajian, J. G., James, N., Woods, D., Chicz-Demet, A., Wilson, A. F., & Sandman, C. A. (1994). Acidosis stimulates beta-endorphin release during exercise. *Journal of Applied Physiology, 77*, 1913–1918.

Teraslinna, P., Partanen T., Koskela, A., Partanen, K., & Oja, P. (1969). Characteristics affecting willingness of executives to participate in an activity program aimed at coronary heart disease prevention. *Journal of Sports Medicine and Physical Fitness, 9*, 224–229.

Warburton, D. E., Nicol, C. W., & Bredin, S. S. (2006). Health benefits of physical activity: The evidence. *Canadian Medical Association Journal, 174*(6), 801–809.

Williams, D. M., Dunsiger, S., Ciccolo, J. T., Lewis, B. A., Albrecht, A. E., & Marcus, B. H. (2008). Acute affective response to a moderate-intensity exercise stimulus predicts physical activity participation 6 and 12 months later. *Psychology Sport Exercise, 9*(3), 231–245.

17 Sports Tourism for People with Disabilities

A Case Study of South Africa

Marie Young, Barry Andrews,
B. Knott, Y. Van der Westhuizen,
Z. Magerman, T. L. Petersen, D. L. Reid,
Angel Mahlalelal, and M. R. Desai

Introduction

Sport and tourism as social constructs have rapidly developed and changed since the 1960s (Higham, 2021). Tournaments originated in France in the 19th century and soon spread across Europe. These tournaments had a sportive aspect. Competitors traveled significant distances to participate in these tournaments that became very popular with spectators. More sporting activities emerged in the latter half of the 19th century and became increasingly mobile: Alpine skiing, yachting, golf, bowling, swimming, and tennis. More recently, national, international, and regional events have attracted significant tourists, participants, and backroom staff to travel for the experience. These events include marathons, various sports "championships," golf events, and horseracing events globally. Examples of such events in South Africa are the Rugby Curry Cup, Cape Epic Cycling event, Comrades Marathon, Two Oceans Marathon, and the Duzi Canoe Marathon. International events include but are not limited to the FIFA World Cup, Cricket World Cup, Rugby World Cup, 7's Rugby World Cup, Olympic Games, Paralympic Games, Special Olympic Games, and many more.

National and international sports tourism events are crucial to economic and social development. Sports tourism is one of the fastest-growing global travel and tourism sectors (United Nations World Tourism Organization [UNWTO], n.d.), worth an estimated $600 billion before the COVID-19 pandemic (World Travel Market [WTM], 2017). South African sport tourism-related activities are rapidly emerging as the primary attraction for international and national travelers worldwide. According to the State of the Industry (SOTI), the number of travelers attending sports events in the United States has increased by more than 10 million since 2015. The UNWTO reports that in 2018, China accounted for 21% of the world's international tourism. The Department of Tourism's annual report suggested that South Africa's tourism base is significant, being one of the world's most popular long-haul destinations. In 2019, South Africa received 1,023 million international arrivals. Sports tourism in South Africa benefits the travel industry immensely as it helps stimulate South Africa's gross domestic product (GDP). All forms of traveling constitute 10% of the world's GDP, accumulating to about $7.6 trillion (World Travel and Tourism Council, 2015).

DOI: 10.4324/9781003476658-21

Sports tourism could be seen as leisure-based travel. It takes people outside their immediate environment to participate in active (active sport tourism) or passive (event sport tourism) recreational or competitive sport, or to appreciate attractions related to physical activities (nostalgia sport tourism). Sport becomes the primary motive for travel, followed by tourism or leisure activities, to enhance the entire experience (Gibson, 2006; Turco et al., 2003). Sports events help shape national and local tourism offerings and transform destinations into desirable event-driven economies (Siyabulela, 2016). Therefore, it can be assumed that sport forms an integral part of developing a country, as sports tourism is often used to re-think and restructure urban and rural communities that are still developing (Nyikana & Tichaawa, 2018).

Sports Tourism and Disability

People with disabilities are an essential target population for the travel industry. Unfortunately, due to historical beliefs, inaccessibility, and a lack of awareness about disability, people with disabilities are still marginalized, globally, in the travel industry and other areas of social life (Carneiro et al., 2021; World Travel and Tourism Council, 2015). It is important to note that sports tourism for people with disabilities begins with knowing the event or country someone wants to visit. The entire process should adopt a universal design approach in the ticketing system, while traveling and when returning home.

Statistics from a quantitative market study in the United States (US) on disability travel showed that there had been an increase in travel. In 2005, people with disabilities took 21 million trips using ground, boat, or air travel, and 26 million trips in 2015 (Open Doors Organisation [ODO], 2015). People with disabilities spend about $17.3 billion annually on their travels and taking into consideration that they might never travel alone, the economic impact doubles to about $34.6 billion (ODO, 2015). Similar results were reported in a study conducted by the European Commission (2014, p. 22) that found "people with special access needs was 17.6 million trips in 2012, of which 7.2 million was taken by people with disabilities and 10.4 million by the elderly population." An Accessible Tourism Market Assessment Study conducted by the South African Department of Tourism (2011, p. 104) indicated that accessible tourism had a significant economic impact on the tourism industry through new business sales with an economic impact of R 12,439 million, new employment creation for 29,249 persons, and "Gross Value Added (GVA) of R5 212 million" that contributes to approximately 0.23% to the GDP of the country. Tourism for people with disabilities proves to be a profitable and growing market for the tourism industry (Shi et al., 2012). In addition to contributing to the economy fiscally, sports tourism promotes new marketing strategies and job opportunities (Wanyonyi et al., 2021).

Hua et al. (2013) claim that sports participation encourages people with disabilities to reclaim their identities and overcome the stigma of disability. Disability is a complex concept defined according to various models (Altman, 2001). Grönvik (2009) endorses this idea, stating that the term "disability" can be defined in various ways, some of which have opposing connotations. The complexity lies in many classifications and models of disabilities such as visual impairments, hearing impairments, cerebral palsy, amputees, spinal cord impairments, and intellectual impairments.

Two traditional models used to define disability are the medical and social models (Smith & Bundon, 2018). Smith and Bundon further note two additional models used to define disability, the social-rational model, and the human rights model of disability. The medical model views disability as a biological product or medical problem that causes limited functioning and seen as deficient. It is believed that the disability could be caused by disease, injury, or health condition, consequently labeling the individual and making him/her feel unvalued (Haegele & Hodge, 2016). The social model implies that society inflicts the disability on the person with an impairment and distinguishes between disability and impairment. It is thus not the body that causes the restriction to functional abilities but the society that causes the restriction (Haegele & Hodge, 2016). The human rights model shares similarities with the social model. However, this model offers a theoretical framework for a disability policy that emphasizes the human dignity of people with disabilities.

The World Health Organization (WHO) (2020) defines disability as an umbrella term encompassing impairments, activity constraints, and participation restrictions. The WHO (2001) uses a classification system called the International Classification of Functioning, Disability and Health. This system is used to identify a person with disabilities' capabilities. It then puts people with similar abilities together to be catered for or compete against each other in a sporting context. This system offers a model to consider multiple impairments related to domains that allow for assessment, goal-setting, and treatment planning.

Disablism, a term used to define discrimination against people with disability, refers to people with disabilities feeling marginalized "via multiple exclusions and deprivations" as it is often "invisible to most, but it is woven deeply into the groundwork of everyday practices and assumptions" (Watermeyer, 2009, p. 148). People with disabilities often feel marginalized by disablism and withdraw from society due to structural inaccessibility, social discrimination, and exclusion (Department of Social Development, 2016). Feelings of exclusion and discrimination surface when they encounter traveling or sporting activities that are inaccessible. While inaccessibility might seem invisible to the fellow non-disabled athlete and traveler, it is generally interwoven into society's assumptions toward people with disabilities (Watermeyer, 2009).

By engaging in an activity, such as sport and tourism, people can experience meaningful practices, promote a sense of belonging and improve well-being (Hammell, 2014). A study completed by Van der Westhuizen (2018) showed that leisure travel for people with disabilities helps redefine their identity. Therefore, there is a great need for accessible sports tourism.

Accessible Sports Tourism

Darcy and Dickson (2009) first coined the term "accessible tourism" – a term used to describe how people with disabilities can travel independently using universally designed products and accessible services. Accessible tourism is defined as tourism that enables people with access requirements (e.g., mobility, vision, hearing, and cognitive dimensions of access) to function independently and with equity and dignity through delivering universally designed tourism products, services, and environments. This definition is inclusive of "all people, including those travelling with children in prams, people with disabilities, and seniors" (Darcy & Dickson, 2009, p. 34).

The main goal for every tourist is to have an enjoyable experience and to be able to participate in as much activity as possible. This is where accessible tourism comes into play. Accessible tourism refers to the notion that ensures all tourist destinations and services are easily accessible to all individuals, including people with disabilities (United Nations, 2022). Many individuals travel in the hopes of experiencing all that a country has to offer. Athletes with disabilities are among those who want to experience the joys and benefits of traveling. However, they are sometimes hesitant due to a lack of information about accessibility, thus uncertain whether the activity or accommodation would cater to their needs.

There are many obstacles of inaccessibility that people and athletes with disabilities have to face to participate in certain tourism activities. A few of the significant barriers they face are transport, accommodation, and inaccessible facilities. South Africa, in general, has a lack of accessible facilities for those with disabilities, which ultimately becomes their demise as this decreases the number of potential tourists it receives. Another barrier on the social side is the lack of knowledge. People are not aware of the needs that those with disabilities have.

Hotel managers, recreation activity managers, and many others must first understand the needs of people with disabilities. Secondly, they must promote a more universal and inclusive environment that permits every tourist, whether they are athletes or not, to participate and make the most of their touristic experiences in the country. Providing universal accessibility is essential for South Africa, especially in creating barrier-free tourism. If the barriers are addressed and removed, it would allow for better access to resources in terms of transportation, accommodation, and tourist attractions to ultimately promote inclusion (Bischoff & Breedt, 2012).

Universal Design for Sports Tourism

Universal design refers to the design of structures that are universally accessible to both people with disabilities and non-disabled people. It enables all to partake in mainstream tourist activities. Having universally designed structures and activities is a fundamental human right as asserted by the United Nations Convention on the rights of people with disabilities, Article 9 (United Nations, 2022). Furthermore, Van der Westhuizen (2018) emphasized that it is the responsibility of the Department of Tourism and the Department of Social Development to remove barriers and identify key role players in universal design and accessibility in mainstream tourist activities.

Ensuring accessibility at sports tourism events will support and establish people with disabilities the human right to travel. Universal Design for sports tourism is not limited to just stadiums but includes transportation, signage, and policies on inclusion and accessibility for all those that could attend the event. The following sections give more detail on universal design for sports tourism development as it pertains to transportation, accommodation, and sports facilities.

Universal Design and Sports Tourism Transportation

South Africa is a popular tourist destination that is reasonably easy to reach. Sport tourism providers aim to ensure that everyone can enjoy tourism, regardless of their physical limits, impairments, or age. Access to transportation is a fundamental human right for all citizens (Morta-Andrews, 2018). Traveling in South Africa is ever growing and is always trying to expand to accommodate people with disabilities. In Cape Town, several transport companies have adapted their vehicles accordingly so that both athletes and tourists, in general, will have access to their transportation needs.

When MyCiTi was developed in 2014, the main goal of the MyCiTi buses was so that everyone, including people with disabilities, had access to some form of transport (MyCiTi, 2021; Wheelchair Travel, 2021b). The MyCiTi buses have wheelchair ramps and handrails on board, allowing athletes and spectators to go sightseeing post-competition. Athletes and spectators can also opt for private transport specifically tailored to their needs. False Bay Care Transport and Tours is another form of transport that is a wheelchair-accessible taxi service that allows practically anyone to travel anywhere in and around Cape Town at any time.

The Dial-A-Ride unique transport service, introduced in May 1998 in response to popular demand for people with disabilities, is funded by the City of Cape Town. People with physical limitations must register to utilize this transportation system, and reservations can be made ahead of time (Hugo et al., 2000). These modes of transport allow people with disabilities to feel a sense of belonging and allow everyone to partake in mainstream activities.

MyCiTi, known as the Integrated Rapid Transit (IRT) system, presently in use within Cape Town, is a case study on the practical application of universal design and accessibility. The IRT system stems from examples globally on the universal design explicitly about infrastructure and construction to improve accessibility for public transport (Smit & Davies, 2012). Wayfinding, known as geographical problem-solving, also refers to methods used by visually impaired individuals. This method enables people with disabilities to participate in mainstream activities as they can freely and securely move between places. The application of Wayfinding is divided into two types, namely orientation and mobility. Orientation is the capability of an individual to observe his or her location in association with the environment. Mobility refers to an individual's ability to travel without harm and identify and evade possible hazards (United States Department of Education, 2001). This tangible wayfinding application/method is available at the Cape Town Stadium. It offers directional information to different regions in the surrounding area, such as car parks and several entries to the stadium.

Additionally, tactile information charts are available at posts in the Green Point Urban Park area, which directs people with visual impairment through the park and the stadium. The design features of the directional signs and information charts are similarly accessible to people with sight due to universally familiar symbols rather than tactile signals that are only made for people who have visual impairments. Tactile wayfinding symbols need to offer vital information in an easy and accessible way to maintain independence for people with disabilities as they reach their destinations quickly and safely (Smit & Davies, 2012).

Considerable efforts have been made to ensure that people with disabilities have accessible transport between Cape Town Stadium and the community, accommodation, and other facilities. The Cape Town Stadium has adhered to universal design, thus being fully accessible to people with disabilities. All facilities, including the sports field, changing rooms, treatment rooms, supporter stands, business lodgings, celebrity areas and offices, are accessible. Additionally, the lighting caters to the specific needs of individuals with visual impairments. Ramps are strategically placed to provide access to the stations for wheelchair users. There is level boarding, which will permit wheelchair entry from the platforms of trunk stations onto buses. All feeder services have mobile ramps to offer on-demand wheelchair entry. There are areas set aside for guarding wheelchairs in both trunk and feeder buses. Buses on the shuttle paths from the airport to the city center and Cape Town Stadium are wheelchair accessible. A barrier-free transport shuttle at one of the public centers in Cape Town will run on all match days. There are also particular induction loops at ticket offices to allow those with a hearing impairment to liaise better with ticket sellers (Barnes, 2010). Special parking has been assigned on DF Malan Drive in Cape Town for those with disabilities who would like to use the stadium transport. Only individuals with a certified disability license from the city can use this specific parking area (SA Rugby, 2019).

Additionally, a company called Access Earth developed an app for people with disabilities to rate the accessibility of structures according to their requirements of accommodation, restaurants, and theatres. The latter significantly contributed to accessible tourism and social change. It promotes easy and accessible traveling, especially for people with disabilities, older adults, and parents with strollers (Access Earth, 2017). To ensure early access to the transport information for people with disabilities, the app can only assist people with disabilities to plan their journey and then partake in events. The app asks a variety of questions such as the width of the door and if it is wide enough for wheelchairs to pass through, if there are wheelchair-accessible toilets available, and if the ramps and showers have railings. This information is made available, accompanied by visuals, to allow people with disabilities to assess the accessibility of the structure concerning their needs prior to visiting.

Universal Design and Sports Tourism Accommodation

There is a lack of accessible facilities for people with disabilities in South Africa. As a result, the government cannot fully benefit from tourism for people with disabilities, and this sector can significantly expand the country's tourism receipts (Bischoff & Breedt, 2012). Finding accessible accommodation, mainly if individuals use wheelchairs or scooters, is one of the challenges that some people with mobility impairments face when going to new places in and around South Africa and abroad (Suursalmi, 2015).

For people with disabilities taking part in accessible tourism, there must be a facility that meets their needs, offering suitable accommodation, adequate technical equipment, appropriate architectural layout, disability awareness from staff, and available support persons (Özcan et al., 2021). Accommodation is a vital part of a tourist's experience. This is where they will recuperate after a long day of sightseeing, so ensuring a comfortable and enjoyable experience is essential. Hotel owners are often unaware of the challenges people with disabilities experience when traveling and, therefore, do not make the appropriate adaptations. Research conducted by Bischoff and Breedt (2012) established that the reception areas at some hotels do not cater to the needs of people with disabilities. The height of the reception counters only suits those who can stand and do not accommodate wheelchair users. There is also a lack of handrails on areas with slopes and an inadequate floor surface, contributing to the increased likelihood of an individual slipping.

However, this is not the case in all the provinces in South Africa. Provinces that are earmarked for major tourism, such as Cape Town, make an effort to accommodate people with disabilities. In South Africa, Cape Town has become one of the most disability-friendly cities and provides accessible accommodation. Several hotels provide roll-in showers for those with mobility impairments and handrails in the shower. This will allow the athlete with a disability to easily shower after a long day of sightseeing in the city (Wheelchair Travel, 2021a).

Universal Design and Sports Tourism Facilities

User-friendly sports facilities must provide accessible services to every member of the public, as accessible facilities are essential to prevent injuries. Organizations that offer accessible sports facilities, explicitly catering to people with disabilities, should market it as such. Specific facilities also have to be available for athletes to achieve awe-inspiring sports performances (Hussein & Yaacob, 2012).

From this understanding, common obstacles to sports participation for people with disabilities can include a lack of perception of incorporating people with different abilities, limited courses and available facilities, poor transportation, and lack of access to resources (DePauw & Gavron, 2005). Given the type of impairment and the variety of disabilities, practical steps in the supply of services and the creation of programs can be taken to assist those with disabilities (Byers et al., 2021).

Considering these obstacles, sports tourism for people with disabilities requires a strategic approach when hosting and developing events. With an influx of people gathering at the event, it becomes increasingly important that these events and venues understand and accommodate everyone, including those with disabilities (Darcy, 2012). Accessibility needs to be a priority as places need to be user-friendly and take both abled and disabled individuals into account. According to Neutens, Schwanen, Witlox, and Maeyer (2010, in Pereira, 2018, p. 47), accessibility can be understood in terms of how easy it is for people to get to certain places and participate in mainstream activities. Having wheelchair accessible bathrooms and stadiums for spectators and athletes is a great example. Not only do wheelchair ramps have to adhere to a 1:12 slope ratio (U.S. Access Board, 2015) with rails, but bathroom stalls and passageways need to be an appropriate width (1 m × 1 m) so that wheelchairs can quickly navigate through.

Creating awareness about accessibility, or the lack thereof, is vital as this can have a knock-on effect and potentially result in tourists not re-visiting the country again, impacting the country's economy. This is especially true in a South African context, where many facilities have not been built to accommodate people with disabilities. Often people with disabilities cannot access facilities due to a lack of universal design. This results in individuals feeling even more excluded and can result in them as tourists not returning since their needs are not being met (Perić et al., 2019). Since the South African government has introduced sports tourism as an essential part of tourism, various initiatives have been launched to contribute to the experience of the athletes and the spectators (Melo & Sobry, 2017). Therefore, it is essential to consider accessibility in tourist attraction sites and stadiums for all people.

Universal Design and Sports Tourism Activities

Many destinations provide accessible activities for people with disabilities. The South African National Parks (2021) have placed great emphasis on the needs of people with disabilities. SANParks have incorporated ramps for those with physical disabilities to access the main facilities, such as the ablution stations at the camping sights. At Kirstenbosch National Botanical Garden, visitor facilities are wheelchair-friendly. However, not all areas in the garden will be accessible (Brand South Africa, 2013). Braille trails are also made available at gardens and nature reserves so that those with visual impairments can participate in these activities. The most popular of its kind is the one at Kirstenbosch Gardens.

Nature tourism is often perceived in terms of interactions and appreciation of nature within a natural environment. It includes factors such as wildlife tourism and adventure tourism and is a big part of the tourism community (Räikkönen et al., 2021). Since nature tourism consists of bird watching, sightseeing, camping, and even hunting, these activities must also cater for people with visual and mobility impairments. There is now a growing demand from people with disabilities who are longing for the same experience as non-disabled sports tourists (Chikuta et al., 2019).

Sports Tourism Events and Disability

Sports tourism events can be categorized by their degree of impact and expenditure of the host city. The highest of these levels is called the mega-event. Mega-events are a significant component of the sports tourism sector. They are defined as events that are ambulatory occasions of a fixed duration that attract many visitors, have an extensive mediated reach, come with high costs, and have significant impacts on the built environment and the population (Müller, 2015). Countries across the world participate in these events, which generates a significant income for the hosting country. Mega-events also attract a lot of media attention to promote future visits from tourists. Therefore, understanding the history of mega-events is essential to comprehend the global trends of sports tourism for people with disabilities. Examples of mega-events in sports include the FIFA World Cup (men and women), IRB Rugby World Cup, F1 Grand Prix, and a limited number of other multi-sport and single sport events.

Growth of Disability Sport Mega-Events

The Olympic and Paralympic Games are the most famous mega-events which occur every four years. The commencement of Olympic and Paralympic Games, dating back to the 1900s, allowed non-disabled individuals and individuals with disabilities to participate and continue doing the sport they love to reap the same benefits.

The history of the Paralympic Games started in 1948 with Dr. Ludwig Guttman, a general practitioner at a hospital in Stoke Mandeville, England. He organized sports competitions as rehabilitation for soldiers with severe spinal cord injuries sustained during the Second World War (Tokyo 2020 Paralympic Games, 2020). Following the competitions and the 1948 Stoke Mandeville Games, Guttmann had a revelation for the future of wheelchair sport beyond the Stoke Mandeville Games. The Paralympic Games directly resulted from that revelation.

It was found that individuals with injuries and disabilities had higher survival rates and longer life expectancy due to the growth of para-sport competitions, such as the Paralympic Games. People with disabilities can now participate in this prestigious event and participate in sports that they love. The International Paralympic Committee (IPC) and the International Disability Alliance launched a global campaign to address the discrimination that disabled people face. The Paralympic Games has helped change societal perspectives, create a more inclusive society, and impact the way people with disabilities viewed their roles in society, including their involvement in sport. The Paralympic Games have progressed from only allowing athletes who use wheelchairs to compete to registering athletes with various disabilities making up the "competition's athlete ordering system."

The Paralympics of Tokyo 2020 were unrecognizable from the last time Tokyo hosted the Games in 1964. During the 1964 games, 378 athletes represented 21 countries and only a small minority of athletes (75) were female. The events comprised nine sports, but only individuals with spinal cord injuries were eligible to compete. As for the 2020 Paralympics, those numbers were almost incomparably more significant. The 2020 Paralympics (hosted in 2021) teams came from approximately 160 nations. They registered approximately 4,400 athletes, with female participants accounting for a record-breaking 40.5% of the total delegation. Athletes competed in 22 sports, with new disciplines continuing to be added.

The viewership of the Paralympic Games has also grown tremendously, with a global television audience of 3.8 billion at the London 2012 Paralympics. The Paralympics and its social media presence are also increasing. For example, at the 2012 London Paralympics, approximately 1.3 million tweets mentioned the term "Paralympic." This diverse, multi-sport, mega-event has grown from one man's idea to use sport as a means for rehabilitation purposes to a global display (Darcy & Legg, 2016).

Another multi-sport mega-event, the Commonwealth Games, takes a different approach to disability and sport by integrating para-sport competition into its overall program. The Gold Coast Commonwealth Games of 2018 was the most significant major sporting event to integrate para-sports into its general sports program (Darcy & Dickson, 2018). The Commonwealth Games has shown an outstanding commitment to expanding para-sports, increasing the para-sport medal events to 38 in 2018 and the number of athletes to more than 300.

The overall responsibility for accessibility and inclusion of the Commonwealth Games athlete experience was integrated into the broader Sustainability Program of the event. The Sustainability Program emphasized the importance of accessibility for the athletes in sports participation and accommodation, and it also included guidelines for the accessibility of event spectators. For example, they stated their aim as "Providing a safe and enjoyable experience for all, ensuring all competition venues will be accessible for people with mobility and other impairments and providing a range of accessible facilities and services" (Darcy & Dickson, 2018, p. 1).

The Invictus Games is a multi-sport mega-event that caters specifically to military service members (both in active service and veterans) with disabilities. First held in 2014, the Invictus Games and its associated foundation aims to use the power of sport to inspire recovery, support rehabilitation, and generate a broader understanding and respect for wounded, injured, and sick service members (Invictus Games Foundation, 2016). The Invictus Games in The Hague, 2022 (postponed from 2020 to 2022), brought together over 500 competitors from 20 nations to compete in a series of adaptive sports (Invictus Games Foundation, 2020).

Accessible Sports Tourism Events in Cape Town

The following two events provide examples of smaller-scale community events that cater to and include people of different abilities.

A significant event that has been held annually in Cape Town since 1981, specifically for people with intellectual disabilities, is the ONE to ONE DAY FAIR. With the support of the Cape Town community, the organization called ONE to ONE has designed this event for people with intellectual disabilities. The goal of the event is to create a peaceful and fun atmosphere for guests to socialize with fellow Cape Town residents and other individuals with disabilities. The entertainment consists of game stalls and live entertainment from the community schools and their students. The fair aims to respect, care for, and promote the rights and welfare of those with an intellectual disability (ONE to ONE, 2018).

Cape Mental Health, hosting the annual Cape Town International Kite Festival, is a registered non-profit mental health organization that provides free community-based services and care for adults and children who have intellectual and psychiatric disabilities. The Kite Festival is a yearly charity and alertness-raising event for Cape Mental Health that inspires and embraces the importance of mental health. As an organization, they believe in the capability of children and adults with mental disabilities to obtain skills, enhance their potential, and live whole and successful lives given the right opportunities, attention, and care. The yearly Kite Festival entices gifted kite-makers and kite fliers from across the globe. The people attending this event will see enormous cartoon characters and other incredible kite designs lifted into the sky and exhilarating stunt kiting. With workshops showing individuals how to design their kites, kite-flying, food stalls, children's rides, and a packed schedule of entertainment, the festival provides fantastic outdoor family fun. The event provides for a terrific cause. All profits will give much-needed mental health services and facilities to adults and children in inadequately resourced communities in the Western Cape (Cape Point Route, 2012).

Conclusion

It will require a concerted effort to create an effective plan for South Africa that ensures universal access. It is vital to raise awareness and understanding of the needs of people with disabilities, especially pertaining to accessible tourism—explicitly regarding accommodation, transport, and activities. Suitable adaptations can provide a pleasant experience for athletes on holiday. Understanding universally accessible infrastructures that maintain set standards (e.g., required dimensions) can help athletes and spectators with disabilities feel included and will result in future tourism opportunities for the city.

Without a doubt, the COVID-19 pandemic affected all aspects of human life, most notably the healthcare system, but also the economy, tourism, education, and sports (Urbański et al., 2021). According to Saarinen and Wall-Reinius (2021), the pandemic raised awareness of the importance of tourism experiences and consumption for individuals and local communities.

In South Africa, all activities ceased on March 5, 2020, when the country was placed on nationwide lockdown due to the COVID-19 pandemic. Businesses, especially in the tourism and leisure industry, had to close, and a relatively large amount of people lost their jobs, leaving the business industry devastated with financial hardship. Movement between provinces was prohibited, and no international traveling was allowed (Young, 2020). These lockdown measures did not only affect South African citizens, but global lockdowns were visible with restricted travel between countries, affecting local and national major events. Examples thereof are the International Olympics 2020 and Paralympics 2020 set to take place in Tokyo, Japan, which was postponed to 2021 due to the severity of the virus. However, no spectators were allowed to prevent the further spread of COVID-19. This meant that the atmosphere was different from previous years because there was no crowd cheering. The spectators play a vital role in sports tourism as they pay for tickets to attend the Olympics in person, which boosts the sports tourism industry fiscally.

Mega events such as the Paralympics, Olympics, and World Cup prove to be a sports tourism drawing card. Therefore, it is imperative to ensure optimal accessibility for athletes and visitors with disabilities. Arguably, tourism in South Africa is not optimally accommodating to travelers with disabilities, and South Africa still faces many challenges of inaccessible accommodation and lack of universal design.

Furthermore, the municipal building control department should enforce the universal design of public spaces by not approving building plans that lack structural accessibility. This will reduce discriminatory barriers and prioritize inclusion of people with disabilities, as pledged by government in the White Paper on the Rights of Person with Disabilities (Department of Social Development, 2016, p. 4).

Providing accessible accommodation, transport, and universally designed facilities creates a successful tourist attraction and greater sports participation. It is vital to ensure that the venue where the competition is taking place is inclusive of all. Further investigation is needed to explore the needs of people with disabilities and make sports tourism more accessible and attractive to this special population that is so easily forgotten.

References

Access Earth. (2017). *We are building the world's largest accessibility database.* Access. earth. Retrieved from http://access.earth/

Altman, B. M. (2001). Disability definitions, models, classification schemes, and applications. In G. L. Albrecht, K. D. Seelman, & M. Bury (Eds.), *Handbook of disability studies* (pp. 97–122). Sage Publication Inc.

Barnes, C. (2010). *Stadium is 'disabled-friendly'.* IOL. Retrieved from https://www.iol.co.za/news/south-africa/stadium-is-disabled-friendly-482344

Bischoff, C. A., & Breedt, T. F. (2012). The need for disabled friendly accommodation in South Africa. *African Journal of Business Management*, 10534–10541. https://doi.org/10.5897/AJBM09.337

Brand South Africa. (2013). *Tourists with disabilities.* Brand South Africa.com. Retrieved from https://www.brandsouthafrica.com/people-culture/people/tourists-with-disabilities

Byers, T., Hayday, E. J., Mason, F., Lunga, P., & Headley, D. (2021). Innovation for positive sustainable legacy from mega sports events: Virtual reality as a tool for social inclusion legacy for Paris 2024 Paralympic Games. *Frontiers in Sports and Active Living, 3*, 10. https://doi.org/10.3389/fspor.2021.625677

Cape Point Route. (2012). *Cape town international Kite Festival.* Cape Point Route. Retrieved from https://www.capepointroute.co.za/blog/activity/cape-town-international-kite-festival/

Carneiro, M. J., Teixeira, L., Eusébio, C., Kastenholz, E., & Moura, A. A. (2021). Use of the internet to plan tourism trips by people with special needs. In C. Eusébio, L. Teixeira, & M. Carneiro (Eds.), *ICT tools and applications for accessible tourism* (pp. 74–95). IGI Global. https://doi.org/10.4018/978-1-7998-6428-8.ch004

Chikuta, O., du Plessis, E., & Saayman, M. (2019). Accessibility expectations of tourists with disabilities in National Parks. *Tourism Planning and Development, 16*(1), 75–92. https://doi.org/10.1080/21568316.2018.1447509

Darcy, S. (2012). Disability, access, and inclusion in the event industry: A call for inclusive event research. *Event Management, 16*(3), 259–265. http://doi.org/10.3727/152599512X13461660017475

Darcy, S., & Dickson, T. J. (2009). A whole-of-life approach to tourism: The case for accessible tourism experiences. *Journal of Hospitality and Tourism Management, 16*(1), 32–44. https://doi.org/10.1375/jhtm.16.1.32

Darcy, S., & Dickson, T. J. (2018). Commonwealth Games have better integrated para-sports, but society needs to catch up. *The Conversation*. Retrieved from https://theconversation.com/commonwealth-games-have-better-integrated-para-sports-but-society-needs-to-catch-up-94491

Darcy, S., & Legg, D. (2016). A brief history of the Paralympic Games: From post-WWII rehabilitation to mega sport event. *The Conversation*. Retrieved from https://theconversation.com/a-brief-history-of-the-paralympic-games-from-post-wwii-rehabilitation-to-mega-sport-event-64809

Department of Social Development (DSD). (2016). *White paper on the rights of persons with disability.* DSD. Retrieved from https://www.gov.za/sites/default/files/gcis_document/201603/39792gon230.pdf

DePauw, K. P., & Gavron, S. J. (2005). *Disability sport*. Human Kinetics.

European Commission. (2014, October). *Economic impact and travel patterns of accessible tourism in Europe – Final Report*. Europa. Retrieved from https://ec.europa.eu/docsroom/documents/7221/attachments/1/translations/en/renditions/pdf

Gibson, H. J. (2006). Sport tourism: Concepts and theories. An introduction. In H. J. Gibson (Ed.), *Sport tourism* (2nd ed., pp. 1–25). Routledge.

Grönvik, L. (2009). Defining disability: Effects of disability concepts on research outcomes. *International Journal of Social Research Methodology, 12*(1), 1–18. https://doi.org/10.1080/13645570701621977

Haegele, J. A., & Hodge, S. (2016). Disability discourse: Overview and critiques of the medical and social models. *Quest, 68*(2), 193–206. https://doi.org/10.1080/00336297.2016.1143849

Hammell, K. R. W. (2014). Belonging, occupation, and human well-being: An exploration: Appartenance, occupation etbien-êtrehumain: Une étude exploratoire. *Canadian Journal of Occupational Therapy, 81*(1), 39–50. https://doi.org/10.1177/0008417413520489

Higham, J. (2021). Sport tourism: A perspective article. *Tourism Review, 76*(1), 64–68. https://doi.org/10.1108/TR-10-2019-0424

Hua, K. P., Ibrahim, I., & Chiu, L. K. (2013). Sport tourism: Physically-disabled sport tourists' orientation. *Procedia-Social and Behavioral Sciences, 91*, 257–269. https://doi.org/10.1016/j.sbspro.2013.08.423

Hugo, J. S., Stanbury, J., & Gooch, J. T. (2000). Demand responsive transport. *19th Southern Africa Transport Conference*. Pretoria, South Africa. Retrieved from https://nailsblaze.com/activism

Hussein, H., & Yaacob, N. M. (2012). Development of accessible design in Malaysia. *Procedia Social and Behavioral Sciences, 68*, 121–133. https://doi.org/10.1016/j.sbspro.2012.12.212

Invictus Games Foundation. (2016). *The Invictus Games Foundation.* Invictus Games Foundation. Retrieved from https://invictusgamesfoundation.org/

Invictus Games Foundation. (2020). *The Hague 202One*. Invictus Games Foundation. Retrieved from https://invictusgamesfoundation.org/games/the-hague-2020/

Melo, R., & Sobry, C. (2017). Introducing sport tourism: New challenges in a globalised world. *European Journal of Tourism Research, 16*, 5–7.

Morta-Andrews, N. (2018). *A case study of transport services for physically disabled citizens in the City of Cape Town.* [Unpublished Masters Thesis], University of the Western Cape.

Müller, M. (2015). What makes an event a mega-event? Definitions and sizes. *Leisure Studies, 34*(6), 627–642. https://doi.org/10.1080/02614367.2014.993333

MyCiTi. (2021). *International disability specialists came Myciti a world leader in universal access 2013-12-04*. MyCiti. Retrieved from https://www.myciti.org.za/en/contact/media-releases/international-disability-specialists-name-MyCiTi-world-leader-in-universal-access/#

Nyikana, S., & Tichaawa, T. M. (2018). Sport tourism as a local economic development enhancer for emerging destinations. *EuroEconomica, 1*(37), 76–89. Retrieved from http://hdl.handle.net/11159/2566

ONE to ONE. (2018). *ONE to ONE Bringing together the intellectually disabled and the caring community: Event held in Cape Town.* One to One. Retrieved from http://www.onetoone.co.za.

Open Doors Organization. (2015). *Market study.* Opendoor SNFP. Retrieved from http://opendoorsnfp.org/wpcontent/uploads/2016/05/ODO-Study-Press-Release-Final.pdf

Özcan, E., Güçhan Topcu, Z., & Arasli, H. (2021). Determinants of travel participation and experiences of wheelchair users traveling to the Bodrum Region: A qualitative study. *International Journal of Environmental Research and Public Health, 18*(5), 2218. https://doi.org/10.3390/ijerph18052218

Pereira, R. H. (2018). Transport legacy of mega-events and the redistribution of accessibility to urban destinations. *Cities, 81*, 45–60. https://doi.org/10.1016/j.cities.2018.03.013

Perić, M., Dragičević, D., & Škorić, S. (2019). Determinants of active sport event tourists' expenditure–the case of mountain bikers and trail runners. *Journal of Sport & Tourism, 23*(1), 19–39. https://doi.org/10.1080/14775085.2019.1623064

Räikkönen, J., Grénman, M., Rouhiainen, H., Honkanen, A., & Sääksjärvi, I. E. (2021). Conceptualizing nature-based science tourism: A case study of Seili Island, Finland. *Journal of Sustainable Tourism, 31*, 1214–1232. https://doi.org/10.1080/09669582.2021.1948553

SA Rugby. (2019). *HSBC Cape Town sevens 2019- What you have to know.* Retrieved August 6, 2021, from https://www.sarugby.co.za/news-features/articles/2019/12/11/hsbc-cape-town-sevens-2019-what-you-have-to-know/.

Saarinen, J., & Wall-Reinius, S. (Eds.). (2021). *Tourism enclaves: Geographies of exclusive spaces in tourism.* Routledge.

Shi, L., Cole, S., & Chancellor, C. (2012). Understanding leisure travel motivations of frequent travellers with acquired mobility impairments. *Tourism Management, 33*(1), 1–83. https://doi.org/10.1016/j.tourman.2011.02.007

Siyabulela, N. (2016). Using sport tourism events as a catalyst for tourism development in the Eastern Cape Province, South Africa. *African Journal of Hospitality, Tourism and Leisure, 5*(3), 1–12. Open Access. Retrieved from http://www.ajhtl.com

Smit, S., & Davies, G. (2012, July 9–12). *The Myciti IRT system in Cape Town: A case study in adhesive tactile way finding.* [Paper presentation]. 31st Southern African Transport Conference 2012. Pretoria, South Africa.

Smith, B., & Bundon, A. (2018). Disability models: Explaining and understanding disability sport in different ways. In I. Brittain & A. Beacom (Eds.), *The palgrave handbook of paralympic studies* (pp. 15–34). Palgrave Macmillan.

South African Department of Sport Tourism. (2011). *Accessible tourism market study 2011.* Resource Cape Town. Retrieved from https://resource.capetown.gov.za/documentcentre/Documents/City%20research%20reports%20and%20review/UAT_Market_Study_2011ND_version.pdf

South African National Parks. (2021). *Parks access descriptions: Summary of accessibility in SANParks for people with mobility difficulties.* Sanparks. Retrieved from https://www.sanparks.org/groups/disabilities/access_parks/access_parks.php

Suursalmi, J. (2015). *Accessible sports tourism: The challenges in travel planning for disabled athletes.* [Bachelor's Thesis], Laurea University of Applied Sciences. Retrieved from https://www.theseus.fi/bitstream/handle/10024/99120/Suursalmi_Joanna.pdf?sequence=1

Tokyo 2020 Paralympic Games. (2020). *About the paralympic games.* Olympics. Retrieved from https://olympics.com/tokyo-2020/en/paralympics/games/paralympic-games-about/

Turco, D. M., Swart, K., Bob, U., & Moodley, V. (2003). Socio-economic impacts of sport tourism in the Durban Unicity, South Africa. *Journal of Sport Tourism, 8*(4), 223–239. https://doi.org/10.1080/1477508032000161537

United Nations. (2022). *United Nations convention on the rights of persons with disabilities.* DESA. Retrieved from https://www.un.org/development/desa/disabilities/convention-on-the-rights-of-persons-with-disabilities/article-9-accessibility.html

United Nations World Tourism Organisation (UNWTO). (n.d.). *Sports tourism.* Retrieved from https://www.unwto.org/sport-tourism

United States Department of Education. (2001). Notice of proposed funding priorities for fiscal years (FYs) 2001–2003 for Two Rehabilitation Research and Training Centers. *Federal Register, 66*(80), 20866–20870.

Urbański, P., Szeliga, Ł., & Tasiemski, T. (2021). Impact of COVID-19 pandemic on athletes with disabilities preparing for the Paralympic Games in Tokyo. *BMC Research Notes, 14*(1), 1–5. https://doi.org/10.1186/s13104-021-05646-0

Van der Westhuizen, Y. (2018). *Exploring the lived experience of leisure travelling for people with disabilities.* [Masters Thesis], University of the Western Cape.

U.S. Access Board. (2015). Guide on ADA standards: Ramps. https://www.access-board.gov/files/ada/guides/ramps.pdf

Wanyonyi, L. N., Njoroge, J. M., & Otieno, R. J. (2021). Challenges and opportunities to sustainable sport tourism events: Insights from an urban host city. *Journal of Tourism, Hospitality and Sports, 55*, 40–55. https://doi.org/10.7176/JTHS/55–06

Watermeyer, B. P. (2009). *Conceptualising psycho-emotional aspects of disablist discrimination and impairment: Towards a psychoanalytically informed disability studies.* [Doctoral dissertation], University of Stellenbosch.

Wheelchair Travel. (2021a). *Accessible hotels in Cape Town.* Wheelchair Travel. https://wheelchairtravel.org/cape-town/hotels/

Wheelchair Travel. (2021b). *Cape Town public transportation.* Wheelchair Travel. https://wheelchairtravel.org/cape-town/public-transportation/

World Health Organization (WHO). (2001). *International classification of functioning, disability and health (ICF).* World Health Organization

World Health Organization (WHO). (2020). *Disabilities.* WHO. Retrieved from https://www.who.int/health-topics/disability

World Travel and Tourism Council. (2015). *Economic impact report 2015.* WTTC. Retrieved from https://www.wttc.org/-/media/files/reports/economic%20impact%20research/regional%202015/world2015.pdf

World Travel Market (WTM). (2017). *GCC countries eye greater share of $600bn global sports tourism industry. Online press report.* WTM. Retrieved from https://hub.wtm.com/gcc-countries-eye-greatershare-of-600bn-global-sports-tourism-industry/

Young, M. E. M. (2020). Leisure pursuits in South Africa as observed during the COVID-19 pandemic. *World Leisure Journal, 62*(4), 331–335. https://doi.org/10.1080/16078055.2020.1825252

Index

Note: **Bold** page numbers refer to tables and *italic* page numbers refer to figures.

For Product Safety Concerns and Information please contact our EU
representative GPSR@taylorandfrancis.com
Taylor & Francis Verlag GmbH, Kaufingerstraße 24, 80331 München, Germany